Nuclear Magnetic Resonance

Edited by Navin Khaneja

Published in London, United Kingdom

IntechOpen

Supporting open minds since 2005

Nuclear Magnetic Resonance
http://dx.doi.org/10.5772/intechopen.74899
Edited by Navin Khaneja

Contributors
Qiu-Liang Wang, Marta C. Corvo, Marcileia Zanatta, Mónica Lopes, Raquel Barrulas, Tiago Paiva, Ana Sofia Ferreira, Gabriele Barbaraci, Mariusz Jaremko, Abdul-Hamid Emwas, Mawadda Alghrably, Samah Al-Harthi, Benjamin Gabriel Poulson, Kacper Szczepski, Kousik Chandra, Zeev Wiesman, Charles Linder, Masatomo Minagawa, Jun Yatabe, Fumio Yoshii, Shin Hasegawa, Nobuhiro Sato, Tomochika Matsuyama, Jürgen M. Schmidt

First published in London, United Kingdom, 2020 by IntechOpen
IntechOpen is the global imprint of INTECHOPEN LIMITED, registered in England and Wales, registration number: 11086078, 7th floor, 10 Lower Thames Street, London, EC3R 6AF, United Kingdom
Printed in Croatia

British Library Cataloguing-in-Publication Data
A catalogue record for this book is available from the British Library

Additional hard and PDF copies can be obtained from orders@intechopen.com

Nuclear Magnetic Resonance
Edited by Navin Khaneja
p. cm.
Print ISBN 978-1-83880-419-0
Online ISBN 978-1-83880-420-6
eBook (PDF) ISBN 978-1-78985-198-4

For EU product safety concerns:
IN TECH d.o.o., Prolaz Marije Krucifikse Kozulić 3, 51000 Rijeka, Croatia, info@intechopen.com or visit our website at intechopen.com.

Meet the editor

Navin Khaneja received his BTech in Electrical Engineering from IIT Kanpur in 1994, followed by his MS and MA in Electrical Engineering and Mathematics from Washington University, St. Louis, in 1997. He earned his PhD from Harvard University in Applied Mathematics in 2000. He is the recipient of the NSF career award, the Sloan fellowship, and the Bessel Prize of the Humboldt Foundation. His research interests are in the areas of control theory, NMR spectroscopy, and quantum control.

Contents

Preface

Nuclear magnetic resonance (NMR) has evolved as a versatile tool in chemistry and biology. The scientific technique is based on the detection of magnetic moments of atomic nuclei arising due to an intrinsic property called spin because of their precession in static magnetic fields. Nuclei are excited by radio frequency (RF) magnetic fields and subsequently their precession is observed by the voltage they induce on an induction coil as they precess. The signal gives valuable information on the precession frequency of nuclei, which depends on the applied magnetic field and the local magnetic fields. At a field of say 14 tesla, protons precess at 600 MHz, carbon at 150 MHz, and nitrogen at 60 MHz. The frequency information is obtained by Fourier transform, which gives a characteristic spectrum. This local field is characteristic of the chemical environment of the nuclei and is termed chemical shift. Chemical shift gives each molecule a fingerprint spectrum with peaks dispersed in the kHz range and helps to identify the molecule from its spectrum. NMR spectroscopy is therefore an important tool in organic chemistry that aids synthetic chemistry. The spectrum of compounds displays characteristic chemical shifts and magnetic couplings between the atomic nuclei. In addition to giving frequency information the NMR signal displays a characteristic decay rate. This decay rate is important in MRI as it helps to provide contrast between the biological tissue being imaged. The decay rate broadens lines in a spectrum of large molecules and makes it difficult to resolve the frequency content of the spectrum.

Modern high-field NMR is able to record the spectrum of large molecules and resolve this spectrum by use of ingenious methods called 2D NMR. These methods rely on carefully tailored RF pulses that correlate frequency of the coupled nuclei by transferring magnetization between coupled spins.

These 2D NMR experiments coupled with relaxation measurements form the basis of the structural analysis of biological molecules, which give information on the dynamics and structure of biological macromolecules. NMR spectroscopy, which started as a tool for the analysis of compounds in organic chemistry, has now matured into a major discipline for the structural and dynamic study of large molecules.

NMR studies are not limited to molecules in solution but are also performed on samples in solid (powder or crystalline) form. These studies on solid-state powders involve spinning the sample to average an anisotropic interaction and obtain a resolved spectrum. These methods have evolved from the study of polymers to the study of biological molecules like membrane proteins.

In this book, we present some of the most exciting developments in the field of NMR: for example, new developments in NMR instrumentation, new magnet technology, RF coil design, the design of novel NMR sensors, and new developments of methods in solution and solid-state NMR. These range from new methods for fast the acquisition of 2D spectrum to NMR studies of molecular interactions in ionic solutions. Solid-state methods for the analysis of polyvinyl chloride and NMR studies of torsion angles in polypeptides are also included.

The book will be a useful reference for practitioners in the field and at the same time will appeal to a broad audience interested in the general area of NMR.

Navin Khaneja
Indian Institute of Technology Bombay,
Mumbai, India

Section 1

NMR Instrumentation

Chapter 1

Hardware of MRI System

Qiuliang Wang

Abstract

Magnetic resonance imaging (MRI) is comprehensively applied in modern medical diagnosis and scientific research for its superb soft-tissue imaging quality and non-radiating characteristics. Main magnet, gradient assembly, and radio-frequency (RF) assembly are main hardware in an MRI system. The hardware performance has direct relationship with the ultimate system overall performance. The development of MRI system toward high magnetic field strength will acquire high signal-to-noise ratio (SNR) and resolution, and meanwhile the manufacture difficulty of main magnet, gradient assembly, and RF assembly will also be significantly elevated. This will make challenges on the design, materials, primitive device, and also the whole machine assembly. This chapter introduces the main hardware of the MRI system and corresponding functions and developments.

Keywords: MRI, superconducting magnet, gradient coil, RF coil

1. Introduction

In the 1950s, it was discovered in the biomedicine field that hydrogen atoms in water molecules can produce nuclear magnetic resonance phenomena [1]. Nuclear magnetic resonance was used to obtain information from the distribution of water molecules in the human body, by which the internal anatomy of the human body could be mapped accurately [2]. After decades of development, MRI has become indispensable medical imaging devices [3]. The influence of MRI on the clinical and life science comes from its unrivaled imaging capabilities, and it can obtain not only clear structural images of the anatomic structure and the organic lesion completely without trauma [4], but also the other physiological information.

In recent years, MRI techniques have developed rapidly, especially toward high-field imaging, such as 7 T, 9.4 T or even higher field strengths [5]. MRI technique development requires an associated performance improvement in the system hardware, which mainly includes the main magnet [6], gradient coil [7], and radio-frequency (RF) coil [8].

2. Main magnet

Early MRI magnet system mainly used ferromagnetic shield structure [9]. The use of large amounts of ferromagnetic shield [10] makes the weight and size of the system relatively large and installation costs of the system high. With the rapid development of magnet technologies, the active shield structure has been successfully developed for the high-field magnet system [11], which greatly reduces the scope of 5 Gauss line. Generally, the superconducting magnet consists of multiple solenoidal coils and shielded coils [12]. The inner solenoid coil is called the primary coil, generally through forward current. The outer solenoid coil is called shield coils, through the reverse current. Open MRI system helps improve patient comfort and expands the scope for the patient [13]. It is easy to achieve a high magnetic field by using combination of iron core and superconducting coil. The cryogenic system is used to keep superconducting wire in a cryogenic environment and ensure safe operation of the superconducting magnet.

In the Institute of Electrical Engineering, Chinese Academy of Sciences, (IEE, CAS), several sets of MRI magnets have been designed or fabricated, including 0.7 T planar whole-body MRI system, 1.5 T cylindrical whole-body MRI system, 7.0 T animal MRI system, and 9.4 T cylindrical whole-body MRI system [14].

Due to structural advantages of the open MRI system, it can be applied to interventional therapy. The shape of magnetic field depends mainly on yoke and pole, and coils provide magnetic source. The magnet system has less superconducting wire, and only 120 L of liquid helium with zero boiling off liquid helium by one GM cryocooler. Magnetic field strength of superconducting open MRI is generally higher than 0.7 T due to a higher uniform magnetic field produced by superconducting coils [15]. **Figure 1** shows a 0.7 T open MRI system designed by IEE, CAS.

A 9.4 T whole body imaging system is developed in IEE, CAS, shown in **Figure 2**. The magnet has a horizontal length of 3.5 m and large warm bore with a diameter of 0.8 m. The magnet system is designed with the minimum cost of wire consumption. By means of the optimal algorithm, the coil sizes and positions are optimized to reduce the coil volume and constrain the magnetic field inhomogeneity. In the optimization, the coil stress and current margins are also constrained to satisfy the coil safety requirements. Fully stable NbTi superconducting wire type WIC (Wire-In-Channel) is employed to wind the coils. Both active and passive quench protection system are employed to protect the magnet from damage during a quench event. The magnet system will reach the final field homogeneity as low as 0.1 ppm (peak to peak) in the central 30 cm DSV and the stability of 0.05 ppm/h.

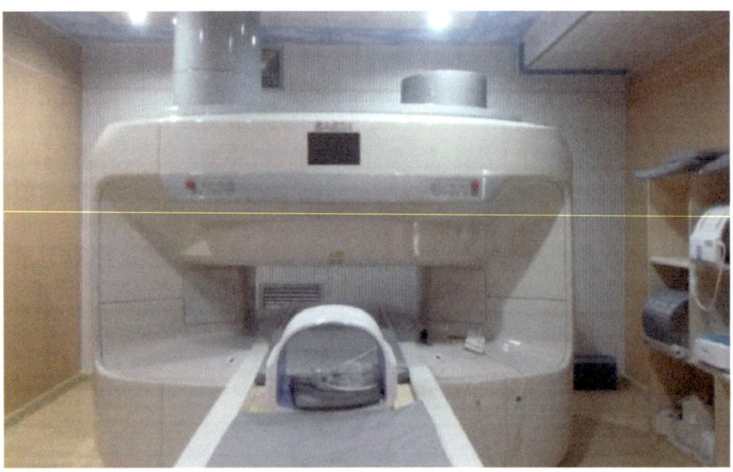

Figure 1.
A 0.7 T whole-body open MRI system.

Figure 2.
A 9.4 T superconducting magnet for whole-body MRI system.

The magnet is cooled by liquid helium bath, and the evaporated helium is condensed by two re-condensing cryocoolers.

3. Gradient coil

A gradient coil set is an important component in a standard MRI scanner which produces linear gradient magnetic fields that are superimposed over a strong uniform magnetic field. The uniform magnetic field is produced by a main magnet, which aligns with the proton precession direction. The superimposed gradient magnetic field slightly changes the proton precession frequency or phase, thus encoding the spatial information of an imaged object in the frequency associated with a position in space [16]. In general, the magnetic field gradient produced by the gradient coils is required to be as linear as possible, and a well-designed gradient coil should also have low inductance, low resistance, high efficiency, etc. [17]. This is especially pertinent in high-field imaging and fast imaging when all the coils' parameters must be highly optimized.

In a gradient assembly, there are three gradient coils, called the x, y, and z coils [18]. **Figure 3** shows a set of actively shielded gradient coils (here the actively shielding gradient coil is a coil pattern containing both the primary coil and shielding coil [19]). The red and blue colors of the gradient coils indicate where the current flows in clockwise and anticlockwise [20]. The three-axis gradient coils are fixed by epoxy resin in an encapsulated gradient assembly [21], as is shown in **Figure 4**. In an integrated gradient assembly, there are also cooling devices and a shim tray installed [22]. The hard epoxy resin largely impedes the vibration of the gradient coils [23], which avoids torsion and deformation of the gradient coils under strong Lorentz force.

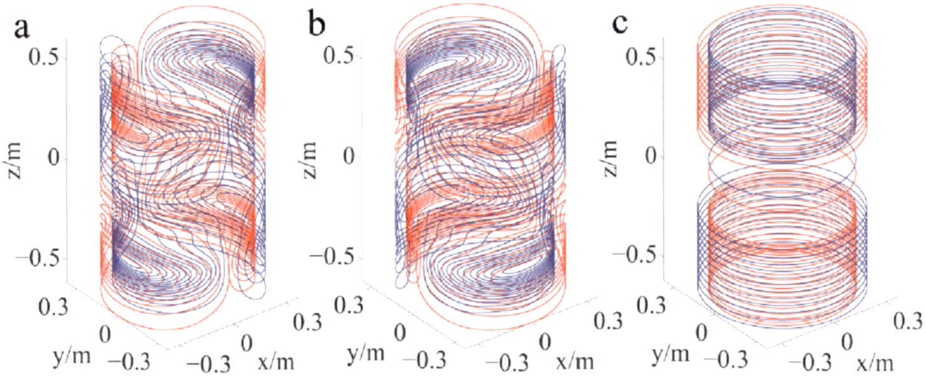

Figure 3.
Actively shielded gradient coils used in an MRI scanner: (a) x gradient coil, (b) y gradient coil, and (c) z gradient coil. The red and blue colors indicate the direction in which the current flows.

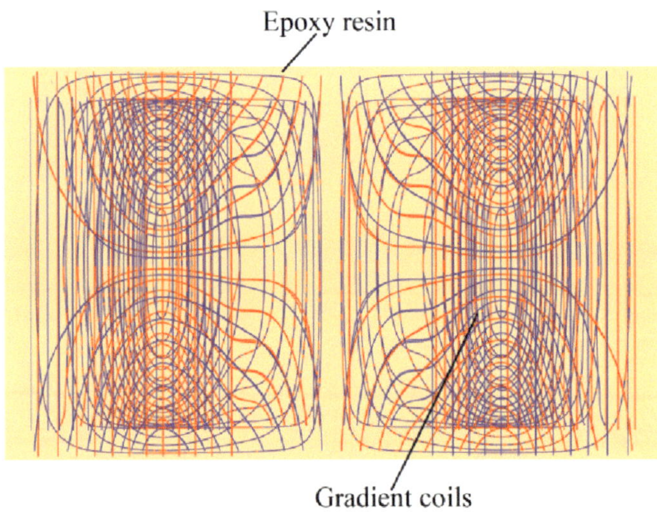

Figure 4.
Illustration of the three-axis gradient coils fixed in the epoxy resin.

4. RF coil

RF coil is the key component of the MRI system, which serves as the transmitter as well as receiver in the formation of the final images [8]. There are various kinds of RF coils. The difference between coils lies in different parts of human body and different field strengths.

According to imaging part of the human body, it can be classified into head coils, body coils, knee coils or foot coils, etc. No matter how many kinds of coils there may be, all the coils can be basically treated as two kinds of coils, namely surface coil and volume coil. For the surface coils [24], the shape of which is usually a circle, which will facilitate the fabrication of coil. Surface coils are often used as receivers, the reason is that the field it produces is inhomogeneous, which is detrimental to the imaging process. But the signal-to-noise ratio (SNR) of the surface coils is higher than volume coils, partly because it can be located closely to the imaging area. Nowadays, surface coils are not used alone to achieve the receiving purpose. A bunch of surface coils [25], which we call loop array, are used for its good performance in receiving as well as transmitting. An illustration of surface coil is shown in **Figure 5**.

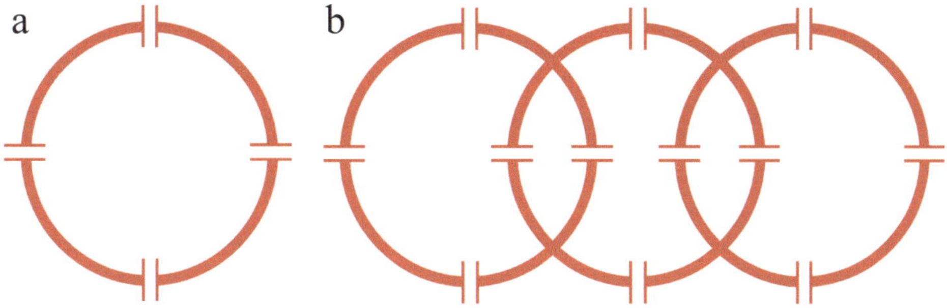

Figure 5.
Diagram of surface coils: (a) single coil and (b) array coil.

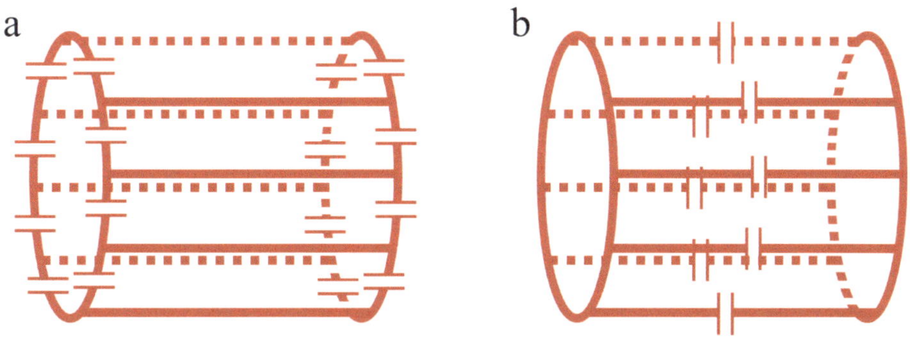

Figure 6.
Diagram of volume coils: (a) high-pass birdcage volume coil and (b) low-pass birdcage volume coil.

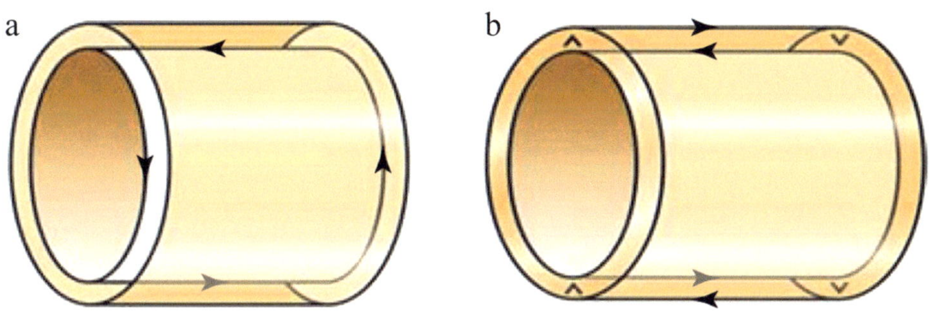

Figure 7.
Sketch map of volume coils: (a) birdcage coil and (b) TEM coil.

Another kind of RF coil is the volume coil, and the most popular volume coil is the birdcage coil [26–29], which is shown in **Figure 6**. A prominent character of birdcage coil is its quadrature excitation strategies both spatially and temporally. The distinctive excitation technology generates a circularly polarized field, which can result in a highly uniform B1 field.

An alternative volume coil named TEM coil [30] maybe used in ultra-high field MRI as illustrated in **Figure 7**. TEM is the most general term for "transmission line," and it can be realized through many kinds of circuit, such as coaxial lines, strip lines, microstrips or waveguides. In birdcage coils, the end rings form the "return path" of the current on the rungs. But in high frequencies, the end rings can be problematic. The inductance and resonant frequency of birdcage coils are limited by the size of the end rings. And the end rings are related to the diameter of the coil. The magnetic field generated by end rings is along with the B0 field, which is nonproductive for the excitation of nucleus. So, in TEM coils, the end rings are replaced by shield to serve as the return path. The inductance and resonant frequency of TEM coil are independent of the diameters of the coil. A length determined TEM coil can be built to arbitrary diameter, without influence on the frequency. So, its use in the ultra-high field imaging is flexible. The TEM coil can be regarded as a toroidal array of many transmission line elements. The impedance of each element can be modified individually. The reactance between each element can be controlled to achieve parallel imaging applications.

Author details

Qiuliang Wang
Institute of Electrical Engineering, Chinese Academy of Sciences, Beijing, China

*Address all correspondence to: wongsm17@mail.iee.ac.cn

References

[1] Gutowsky H. Nuclear magnetic resonance. Annual Review of Physical Chemistry. 1954;**5**:333-356

[2] Brown RW, Cheng YN, Haacke EM, et al. Magnetic Resonance Imaging Physical Principles and Sequence Design. Hoboken, New Jersey: John Wiley and Sons, Ltd; 2014

[3] Lakrimi M, Thomas AM, Hutton G, et al. The principles and evolution of magnetic resonance imaging. Journal of Physics: Conference Series. IOP Publishing. 2011;**286**:012-016

[4] Wang Q, Dai Y, Zhao B, et al. Development of high magnetic field superconducting magnet technology and applications in China. Cryogenics. 2007;**31**:364-379

[5] Robitaille PM, Berliner L. Ultra-High Field Magnetic Resonance Imaging. New York, NY, USA: Springer, LLC; 2006

[6] Wang Q. Practical Design of Magnetostatic Structure Using Numerical Methods. Singapore: John Wiley & Sons; 2013

[7] Siebold H. Gradient field coils for MR imaging with high spectral purity. IEEE Transactions on Magnetics. 1990;**26**:897-900

[8] Vaughan JT, Griffiths JR. RF Coils for MRI. Chichester, West Sussex: John Wiley and Sons, Ltd; 2012

[9] Wang Q, Wang H, Zheng J, et al. Open MRI magnet with iron rings correcting the Lorentz force and field quality. IEEE Transactions on Applied Superconductivity. 2014;**24**:1-5

[10] Yandong Y, Dai Y, Wang H, et al. Geometrical optimization of iron shield for high field superconducting magnetics in MRI system. Chinese Journal of Low Temperature Physics. 2011;**33**:372-376

[11] Gabrielse G, Tan J. Self-shielding superconducting solenoid systems. Journal of Applied Physics. 1988;**63**:5143-5148

[12] Wang Q, Dai Y, Zhao B, et al. A superconducting magnetic system for whole-body metabolism imaging. IEEE Transactions on Applied Superconductivity. 2012;**22**:4400905-4400905

[13] Wang QL, Xu G, Dai Y, et al. Design of open high magnetic field MRI superconducting magnet with continuous current and genetic algorithm method. IEEE Transactions on Applied Superconductivity. 2009;**19**:2289-2292

[14] Dai Y, Wang Q, Wang C, et al. Structural design of a 9.4 T whole-body MRI superconducting magnet. IEEE Transactions on Applied Superconductivity. 2012;**22**:4900404-4900404

[15] Ni Z, Hu G, Li L, Yu G, Wang Q, Yan L. Globally optimal algorithm for design of 0.7 T actively shielded whole-body open MRI superconducting magnet system. IEEE Transactions on Applied Superconductivity. 2012;**23**:4401104-4401104

[16] Pykett IL, Newhouse JH, Buonanno FS, Brady TJ, Goldman MR, Kistler JP, et al. Principles of nuclear magnetic resonance imaging. Radiology. 1982;**143**:157-168

[17] Tuner R. Gradient coil design: A review of methods. Magnetic Resonance Imaging. 1993;**11**:903-920

[18] Jin JM. Electromagnetic Analysis and Design in Magnetic Resonance Imaging. Boca Raton: CRC Press; 1999

[19] Sanchez H, Liu F, Trakic A, et al. A simple relationship for high

efficiency-gradient uniformity tradeoff in multilayer asymmetric gradient coils for magnetic resonance imaging. IEEE Transactions on Magnetics. 2007;**43**:523-532

[20] Lopez HS, Liu F, Poole M, et al. Equivalent magnetization current method applied to the design of gradient coils for magnetic resonance imaging. IEEE Transactions on Magnetics. 2009;**45**:767-775

[21] Hidalgo-Tobon SS. Theory of gradient coil design methods for magnetic resonance imaging. Concepts in Magnetic Resonance Part A. 2010;**36**:223-242

[22] You XF, Yang WH, Song T, et al. Asymmetric gradient coil design by numerical approach for MRI brain imaging. IEEE Transactions on Applied Superconductivity. 2012;**22**:4401904-4401904

[23] Suits BH, Wilken DH. Improving magnetic field gradient coils for NMR imaging. Journal of Physics. 1989;**22**:565-573

[24] Ackerman JJ, Grove TH, Wong GG, et al. Mapping of metabolites in whole animals by 31P NMR using surface coils. Nature. 1980;**283**:167-170

[25] Roemer PB, Edelstein WA, Hayes CE, et al. The NMR phased array. Magnetic Resonance in Medicine. 1990;**16**:192-225

[26] Hayes CE, Edelstein WA, Schenck JF, et al. An efficient, highly homogeneous radiofrequency coil for whole-body NMR imaging at 1.5 T. Journal of Magnetic Resonance. 1985;**63**:622-628

[27] Leifer MC. Theory of the quadrature elliptic birdcage coil. Magnetic Resonance in Medicine. 1997;**38**:726-732

[28] Vaughan JT, Hetherington HP, Otu JO, et al. High frequency volume coils for NMR imaging and spectroscopy. Magnetic Resonance in Medicine. 1994;**32**:206-218

[29] Tropp J. The theory of the birdcage resonator. Journal of Magnetic Resonance. 1989;**82**:51-62

[30] Pascone RJ, Garcia BJ, Fitzgerald TM, et al. Generalized electrical analysis of low pass and high pass birdcage resonators. Magnetic Resonance Imaging. 1991;**9**:395-408

Chapter 2

Control Flow Strategy in a Receiver Coil for Nuclear Magnetic Resonance for Imaging

Gabriele Barbaraci

Abstract

A mathematical discussion is introduced to describe the receiver coil characterizing the NMRI system starting from a general shape of the conductor. A set of different inductance calculations have been introduced varying the shape of the conductor. The inductance calculation led to a general expression of the magnetic field of a single coil characterized by a rectangular shape. A dynamic model of the receiver coil has been developed to represent the natural frequencies that characterize the operational bandwidth. A nonstationary control strategy is implemented to make a real-time changing of the operational bandwidth. The frequency response of the coil generates the necessary conditional expression in order to let the peak of resonance move to a desired value of frequency.

Keywords: magnetic resonance, inductance, magnetic field, transfer function

1. Introduction

For a NMRI system to work, a set of operations must be completed in a time order. The first operation is characterized by choosing the point of the human body that must be analyzed; determining the resonance magnetic field which must be associated to that point; the RF magnetic field pulsing sequence; phase encoding process; and data extraction processed by a digital microprocessor proving a high computational burden. In this chapter, one step characterizing the NMRI system that represents the most important physical effect involving a dynamic aspect which is the transmitter coil will be discussed. In the NMRI systems, it is strictly necessary to have a transmitter coil and a receiver coil, respectively, to excite the magnetization vector inside the human body and transform in electrical current signal the magnetic field variation coming from the magnetization vector. The modern NMRI exhibit a more compact design characterized by associating to one coil the task to generate the RF pulse and receive the signal. This is the reason why in the literature and in the real operating NMRI, the use of a term that refers to a coil performing both tasks called trans-receiver is very common. In this chapter, a model of receiver coil will be studied by focusing the attention in the receiving task because of the importance that is related to the image's quality. The receiver coils can be considered as a filter that allows to capture signal's frequency of interest. As a filter, the receiver coil can be a low-pass, high-pass, or band-pass filter. The challenge of all NMRI manufacturers and of most of the scientific researchers is to

IntechOpen

improve the image's quality because of its importance in early-stage diagnosis and make visible the blood vessels. The improvement of image's quality can occur by eliminating the noise and at the same time having a wider spectrum. The images' quality can be also improved by increasing the signal-to-noise ratio (SNR) as a variable of a design which the receiver coil is related to. The SNR calculation in magnetic resonance imaging (MRI) coil arrays is a powerful tool in the development of new coil arrays. A proposed method describes a model that allows the calculation of the absolute SNR values of arbitrary coil arrays [1]. Another sophisticated strategy is to implement a filter that can estimate the current signal no matter what the noise is. This filter, called also estimator, is commonly used in the control system theory to capture those values of state's vector, whereas a sensor is of difficult implementation. The most common estimator provides a very accurate reconstruction of the missing signal or the signal affected by an amount of noise characterized by a certain variance. Typically, the *Kalman* filter and *Luenberger* observer can provide an excellent reconstruction of the signal [2]. The receiver coils might have different shapes and dimensions according to the specific part of the bod that must be analyzed. The shape can be also optimized to reduce the noise or to minimize the magnetic field coming from the RF pulse produced by the transmitter coil. The shape can be planar which is distributed as an array [1] or a solenoid shape [3, 4]. In high magnetic field pulse NMRI, the risk of inhomogeneity is very high; this leads to an alteration of the current signal carried out from the receiver coils. A classical finite difference time domain (FDTD) coupled with a transverse electromagnetic resonator provided the use of high magnetic field pulse in image extraction [5]. The dimension of a receiver coil may vary from order of magnitude like the NMRI system up to small dimension like a human finger. In some cases, micro-coils are used for NMR microscopy for the possibility to capture a signal not affected by noise and at the same time have a large bandwidth of mode signal [6–9]. A use of a simulation has characterized the design of NMRI system since the computers with strong computational capabilities have been built. The use of a simulator has also characterized the study of NMRI receiver coil and in some cases has constituted a valid tool to investigate on the coil performance over a range of different tuning, allowing the engineers in saving an enormous amount of time [10]. A good frequency resolution has been reached by using an optimization algorithm of strip-line chip [11]. This chip has been built up in a silicon substrate, and its geometry was modeled in respect of RF-homogeneity, sensitivity, and spectral resolution. This model allows to achieve a resolution of 0.7 *Hz*, leading to more visibility of all the harmonic during the DFT analysis. Another shape of receiver coils is the so-called birdcage coil [2, 3] that works also as a transmitter coil. This shape is characterized by a circular loop array having specific axial length and diameter approximately equal to the space where the patient is introduced. An interesting technique in improving the performance of a receiver coil is to use a sliding tuning of the ring characterizing the birdcage coil configuration. The sliding tuning allows to vary the capacity's values and so also the operative bandwidth [12]. The modern technique in NMRI receiver coil is the use of stacked resonator. This configuration is characterized by a multiple-connection of receiver coil one to another to cover the entire body's surface. This allows to maximize the magnetic energy the human body releases during the free induction decay (FID) [13, 14]. A method that develops an optimization of RF coil is based on the algorithm called method of moment that leads to a set of different geometry. This method allows to include in the design the desired bandwidth operation of the signal coming out from the human body. This is a very good strategy in designing RF coils, but it does not allow to vary in real time the bandwidth and so be adapted in several operational conditions [15, 16]. The value of current covers an extreme importance in designing RF coils since the power of the signal and so the image quality depend on it. A free space time-harmonic electromagnetic *Green*'s functions and de-emphasized magnetic field pulse target fields are used to calculate the current density on the coil cylinder to avoid an overemphasis of some areas in the human body that do not allow to see in a clear way the details [17]. One of the most used techniques is to increase the value of uniform magnetic field since it provides a higher resonance frequency and so a higher energy transmitted from the body to the antennas. The modern values of NMRI system can reach a magnetic field equal to 7 T [18]. It has been demonstrated that a high value of magnetic field might not be a good solution for humans having an age up to 5 years old since the magnetic field can cause nerve contraction or temporary disease. This is the reason why a suitable RF coil design is the most accepted strategy particularly in those cases where it is necessary to increase the details without exceeding a certain value of magnetic field. The RF coils are typically shaped according to the surface where they must be located; for example, there can be one for the head, hand, and leg [19, 20]. An innovative method in the analysis of RF coils is represented by the application of the boundary element method with regularization technique. This method represents an effective approach for solving the electromagnetic forward problem and includes biplanar transverse gradient coils and RF phased array coils [21, 22]. With the consideration of the practical engineering requirements, physical constraints such as wire intervals are transformed into mathematical constraints and formulated into BEM equations. The examples demonstrate that the proposed method is efficient and flexible for the design of MRI coils with arbitrary geometries and engineering constraints. The modern NMRI system has also the possibility to work with low and high magnetic field values to make them more adapted to a wide range of clinical cases. In order to increase the quality of the images, there has to be an optimization to two different aspects or rather the transmission mode where RF coils must be able to produce a uniform magnetic field in the volume of interest so that the nuclei can be properly excited; the receiving mode, a SNR, is needed, and the coil must be able to collect the signal emitted by the nuclei with better sensitivity throughout the volume of interest. A strategy to increase the SNR has been proposed in [23] by dropping the temperature up to 77 K.

2. RF coils

In this section the RF coils are introduced. The RF coils have usually two tasks: produce the magnetic field pulse and receive the signal from the magnetization vector. As mentioned before, in this chapter, the receiving performance of the RF coil will be discussed. The RF coils must be able to provide high SNR and maximize the power of electrical current signal generated by the magnetic flow produced by the rotation of magnetization vector during the FID. The noise which they undergo is generated because of the inductance phenomena between one loop and another. The inductance is a function of current and magnetic flow linking the winding wire characterizing the coil. The cross section of wires can assume two different shapes, circular or square [24], and they will be discussed in this section by using mathematical analysis that allows a comparison between them.

2.1 Transmitter and receiver coils

The excitation of hydrogen nuclei placed inside the human body is produced by a magnetic field that magnetizes the nuclei along the direction of the magnetic field $\vec{B}_{0z}$.

The field $\vec{B}_{0z}$ is the magnetic field directed towards the direction feet to head of the patient, determining the magnetic polarization of the human body. The field $\vec{B}_{0z}$ is produced by the uniform axial magnetic field and by the gradient coil. The transmitter coil produces a magnetic field called $\vec{B}_1(t)$, directed along the transversal direction to the magnetic field $\vec{B}_{0z}$, and is a function of time because of the impulse characteristic that has a function $Sinc(\omega t)$ where ω is the *Larmor* frequency. The resonance occurs because frequency ω of the impulse is the same of the oscillation frequency of hydrogen protons. To capture the oscillations, the receiver coil is introduced and characterized by several electrical parameters such as resistance, inductance, and capacitive components associated also to supplementary circuits providing current circulation that works as filter. The resistances are represented by copper wire or by the resistors; the inductance is produced by the mutual and auto effect of coils crossed by current. The capacitors are introduced in the receiver coil as common electrical components with a specific value according to the desired operative bandwidth. The most difficult part in designing a receiver coil is to model the inductance according to the shape of the coil. The inductance modeling and the knowledge of the operative bandwidth will lead to the dynamic mathematical model of the receiver coil from where the control system will be built up.

2.1.1 Self-inductance of and finite length wire

The phenomenon of self-induction occurs every time there is the presence of a magnetic flux that is concatenated to the circuit where the current that generates that flow flows. This flow generates an EMF which opposes the causes that generated it causing alteration of voltage along the branches of the circuit. The case in which the wire is a conductor having a diameter $2r_w$ will be analyzed to then study the case of a conductor with a rectangular amplitude cross section w. The hypothesis that is made in this study is that the current is uniformly distributed over the whole cross section of the conductor; this allows us to treat the problem as the one inherent to a filament as shown in **Figure 1**.

In **Figure 1** the wire has a finite length l where the surface for the calculation of the flow is given by one side equal to the length of the wire, while the other is of infinite value. Recall that inductance L_Φ is given by the ratio between the flow produced by the magnetic field produced by the current flowing in that length wire l and the current that generates that same field as described by (1):

$$L_\Phi = \frac{\Phi\left(\vec{B}\right)}{i_{\vec{B}}} \tag{1}$$

To develop the expression of self-inductance, the expression of the magnetic field is developed in two variables describing the distance transverse to the wire and that along the wire itself as shown in **Figure 2**.

From *Biot-Savart*'s law, it has:

$$dB = \frac{\mu_0 I}{4\pi R^2} \sin(\vartheta) dz \tag{2}$$

The distance R is calculated from the wire element; the point can be represented as:

$$R = \sqrt{(Z - z)^2 + r^2} \tag{3}$$

where

$$\sin(\vartheta) = \frac{r}{R} \tag{4}$$

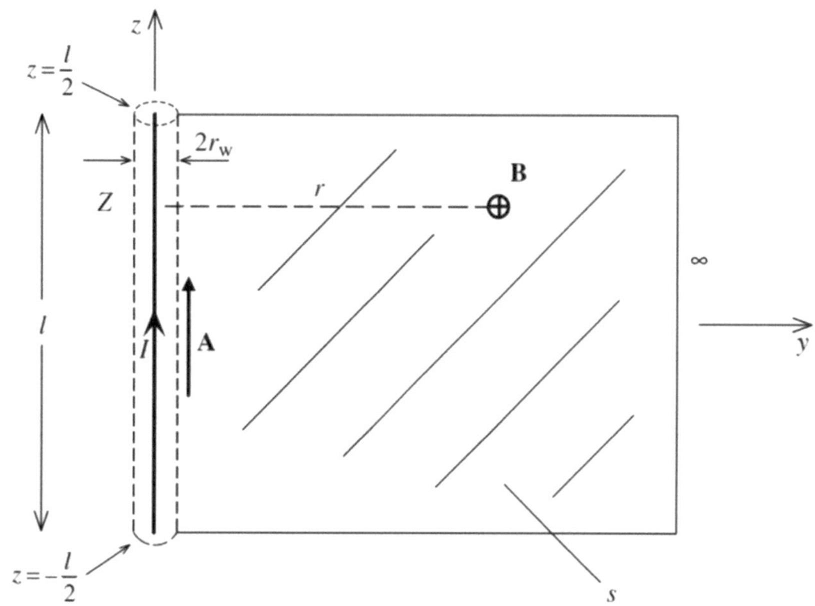

Figure 1.
Self-inductance case study for a finite length wire.

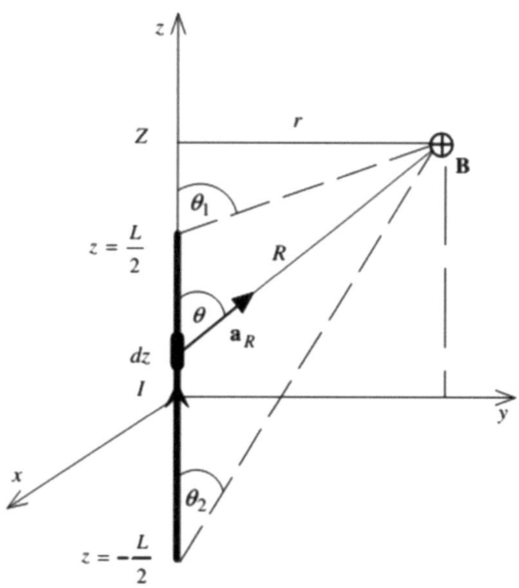

Figure 2.
Scheme for magnetic field B.

Therefore, the magnetic field produced by a wire crossed by a certain current and having a total length equal to L in all points of space is given as:

$$\left|\vec{B}\right| = \frac{\mu_0 I}{4\pi}\int_{-\frac{L}{2}}^{\frac{L}{2}} \frac{r}{\left[(Z-z)^2 + r^2\right]^{3/2}} dz \tag{5}$$

By changing variables as $Z - z = \lambda,\ d\lambda = -dz$, it has the integral (6):

$$\left|\vec{B}\right| = \frac{\mu_0 I r}{4\pi}\int_{Z-\frac{L}{2}}^{Z+\frac{L}{2}} \frac{1}{\left(\lambda^2 + r^2\right)^{3/2}} d\lambda \tag{6}$$

from where the expression of magnetic field:

$$\vec{B}(r,Z) = \frac{\mu_0 I}{4\pi r}\left(\frac{Z+L/2}{\sqrt{r^2 + (L/2+Z)^2}} - \frac{Z-L/2}{\sqrt{r^2 + (L/2-Z)^2}}\right) u_\vartheta \tag{7}$$

This vector generates a flow through the surface shown in **Figure 1** as:

$$\Phi\left(\vec{B}\right) = \int_{r=r_w}^{\infty}\int_{Z=-l/2}^{l/2} \vec{B}\ dZ\ dr \tag{8}$$

that returns the expression in (9):

$$\Phi\left(\vec{B}\right) = \frac{\mu_0 I}{2\pi} L\left[\log_e\left(\frac{L}{r_w} + \sqrt{\left(\frac{L}{r_w}\right)^2 + 1}\right) - \sqrt{1 + \left(\frac{r_w}{L}\right)^2} + \frac{r_w}{L}\right]_{r=r_w}^{r\to\infty} \tag{9}$$

from where the auto-inductance is:

$$L_\Phi = \frac{\mu_0}{2\pi} L\left[\log_e\left(\frac{L}{r_w} + \sqrt{\left(\frac{L}{r_w}\right)^2 + 1}\right) - \sqrt{1 + \left(\frac{r_w}{L}\right)^2} + \frac{r_w}{L}\right] \tag{10}$$

By using the property, it is defined as:

$$\sinh^{-1}\left(\frac{L}{r_w}\right) = \log_e\left(\frac{L}{r_w} + \sqrt{\left(\frac{L}{r_w}\right)^2 + 1}\right) \tag{11}$$

Eq. (10) is defined as:

$$L_\Phi = \frac{\mu_0}{2\pi} L\left[\sinh^{-1}\left(\frac{L}{r_w}\right) - \sqrt{1 + \left(\frac{r_w}{L}\right)^2} + \frac{r_w}{L}\right] \tag{12}$$

Eq. (10) can be simplified in the hypothesis that $L \gg r_w$:

$$L_{\Phi approx.} = \frac{\mu_0}{2\pi} L\left(Log_e\frac{2L}{r_w} - 1\right) \tag{13}$$

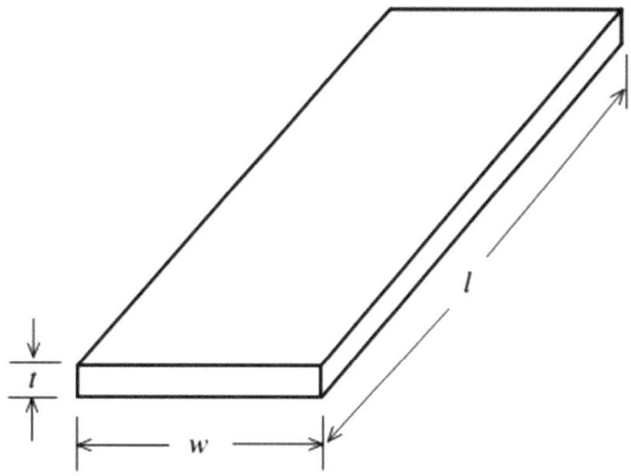

Figure 3.
Rectangular cross-section wire.

In case the wire has a rectangle cross section with $w \gg t$ as shown in **Figure 3**, the expression of self-inductance is [24]:

$$L_\Phi = \frac{\mu_0}{2\pi}\frac{1}{w^2}\left[Lw^2 Log_e\left(\frac{L}{w}+\sqrt{\left(\frac{L}{w}\right)^2+1}\right) + L^2 w\, Log_e\left(\frac{w}{L}+\sqrt{\left(\frac{w}{L}\right)^2+1}\right) + \frac{1}{3}\left(L^3+w^3\right) - \frac{1}{3}\left(L^2+w^2\right)^{3/2} \right] \tag{14}$$

that, when simplified according to the assumption $w \ll L$, returns [3]:

$$L_{\Phi approx.} = \frac{\mu_0}{2\pi} L\left(Log_e \frac{2L}{w} + \frac{1}{2} \right) \tag{15}$$

In **Figures 4** and **5**, a comparison between the exact expression of inductance and the approximated one, respectively, for round cross-section wire and rectangle is shown:

In the same figures, it is possible to see how the values of the characteristic cross section higher than 0.01 m of the approximated relation deviate from the exact one tending to reach a significant difference. All curves shown in **Figures 4** and **5** are characterized also by a decreasing of inductance by increasing the characteristic dimension of the cross section. This happens because in the expression of inductance, the ratio is between the magnetic field flow and current, so by increasing the distance from the wire, the magnitude of the magnetic field decreases and the flow as well:

2.1.2 Mutual inductance between two parallel finite length wires

For the calculation of the mutual inductance, it is assumed that the parallel wires have the same length and are aligned as shown in **Figure 6**.

The calculation remains the same with the only difference that the integration interval changes as $y = [d + r_w, \infty)$. Eq. (16) shows the general expression of the mutual inductance [24]:

$$M_\Phi = \frac{\mu_0}{2\pi} L\left[Log_e\left(\frac{L}{d+r_w}+\sqrt{\left(\frac{L}{d+r_w}\right)^2+1}\right) - \sqrt{1+\left(\frac{d+r_w}{L}\right)^2} + \frac{d+r_w}{L} \right] \tag{16}$$

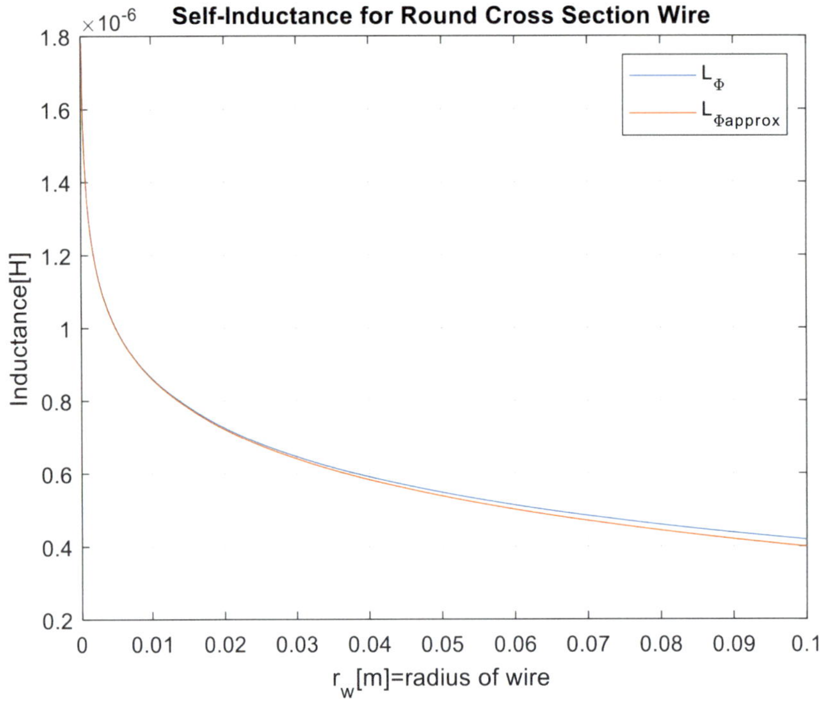

Figure 4.
Partial self-inductance comparison for round wire cross section.

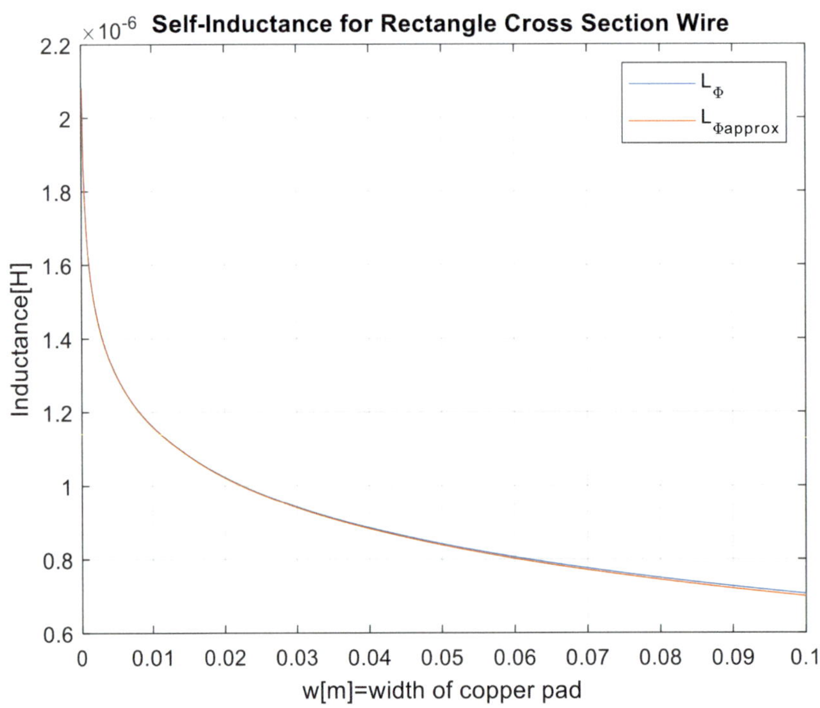

Figure 5.
Partial self-inductance for rectangle wire cross section.

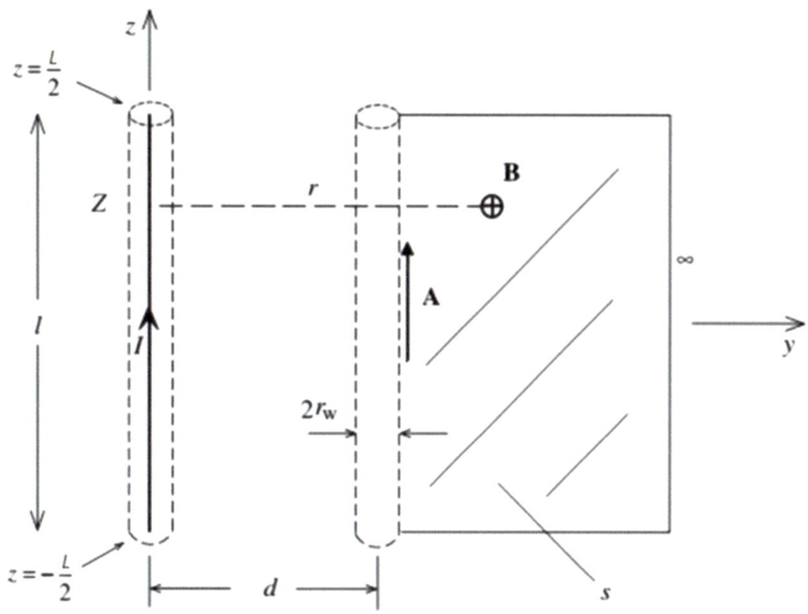

Figure 6.
Scheme to calculate the partial mutual inductance for two parallel finite length wires.

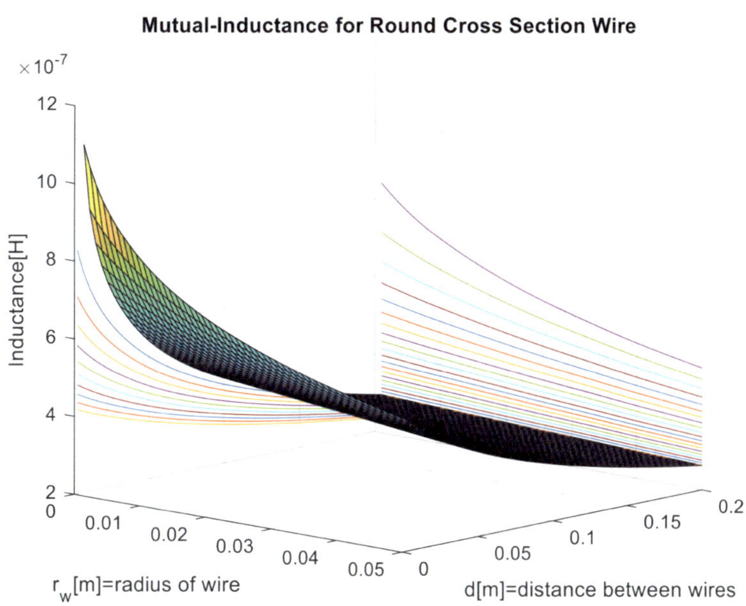

Figure 7.
Partial mutual inductance for round cross-section wires.

that based on the condition $d \gg r_w$ returns the approximated expression:

$$M_{\Phi approx.} = \frac{\mu_0}{2\pi} L \left[Log_e \left(\frac{L}{d} + \sqrt{\left(\frac{L}{d}\right)^2 + 1} \right) - \sqrt{1 + \left(\frac{d}{L}\right)^2} + \frac{d}{L} \right] \tag{17}$$

In **Figure 7**, the variation of Eq. (2.16) as a function of d and r_w has been shown. The inductance decreases with an increasing of the geometrical parameters characterizing the dimension of the cross section of wire and their mutual distance. In the same figure, the inductance as a function of wire's ray shows a slower rate than the

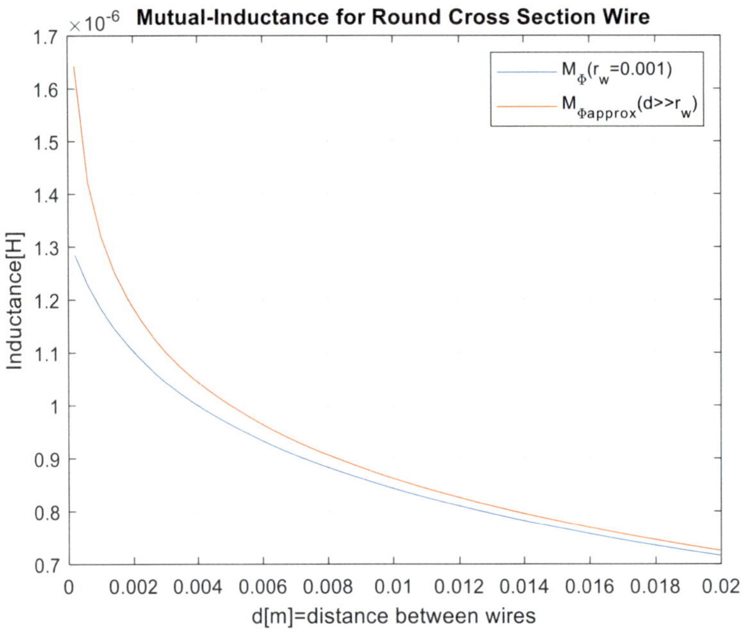

Figure 8.
Partial mutual inductance comparison.

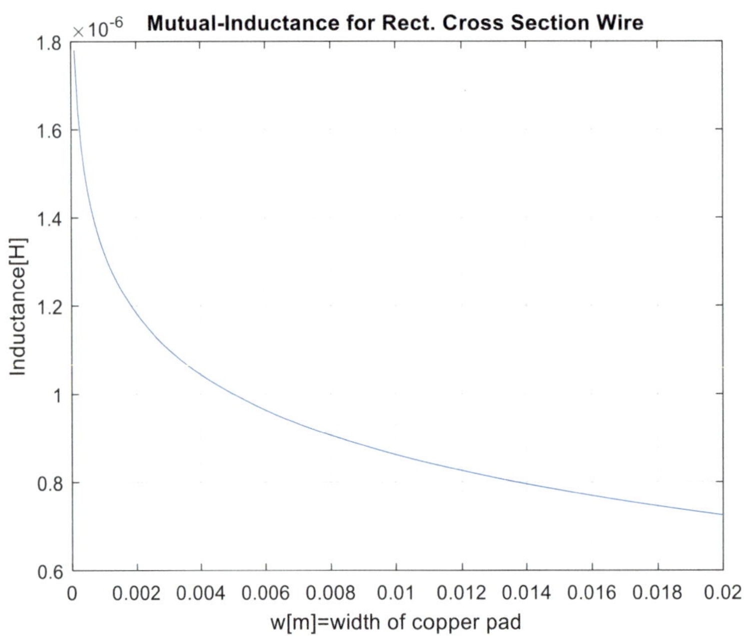

Figure 9.
Partial mutual inductance for rectangular cross-section wires.

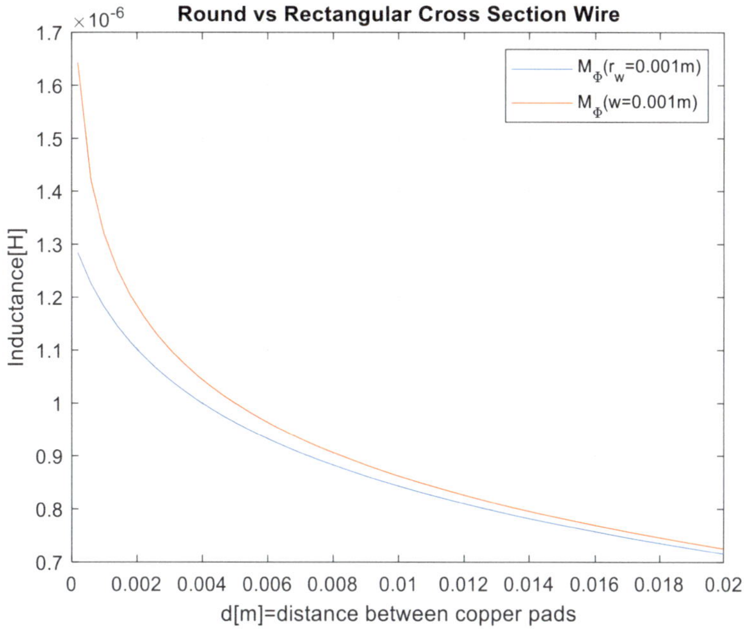

Figure 10.
Partial mutual inductance comparison.

inductance as a function of the distance. In **Figure 8**, the comparison between the exact and approximated expression of the inductances has been shown. For $d \cong r_w$ the difference is significant. In the same figure, it is clear also that an increasing of d equal to $d = 10 \times r_w$ and the two expressions tend to become coincident validating the assumption $d \gg r_w$.

In case of rectangular cross-section wire with $w \gg t$, the expression is shown in Eq. (18) obtained by imposing $d = w$ in Eq. (16) [24].

$$M_{\Phi lin} = \frac{\mu_0}{2\pi} L \left[Log_e \left(\frac{L}{d+w} + \sqrt{\left(\frac{L}{d+w}\right)^2 + 1} \right) - \sqrt{1 + \left(\frac{d+w}{L}\right)^2} + \frac{d+w}{L} \right] \tag{18}$$

In **Figure 9** the pattern of Eq. (18) describing the inductance produced by a rectangular cross section wire has been shown. The inductance decreases by increasing the width of pads keeping the same fashion of the magnetic field pattern of a partial self-inductance shown in **Figure 5**. Moreover, in **Figure 10** a comparison between the round cross section and the circular one has been reported. In the same figure, it is clear how the rectangular cross section exhibits a higher value of the inductance for very low mutual distance between the pads. Again, in **Figure 9** the two patterns tend to become coincident by increasing the mutual distance. This happens because for infinite distance the two wires can be represented as a filament no matter what their own shapes are.

In this section the self-inductance and mutual inductance of finite length wires crossed by a uniformly distributed current and characterized by different geometrical parameters have been calculated. In the next section, the partial mutual inductance for parallel wires located in different axial positions will be shown.

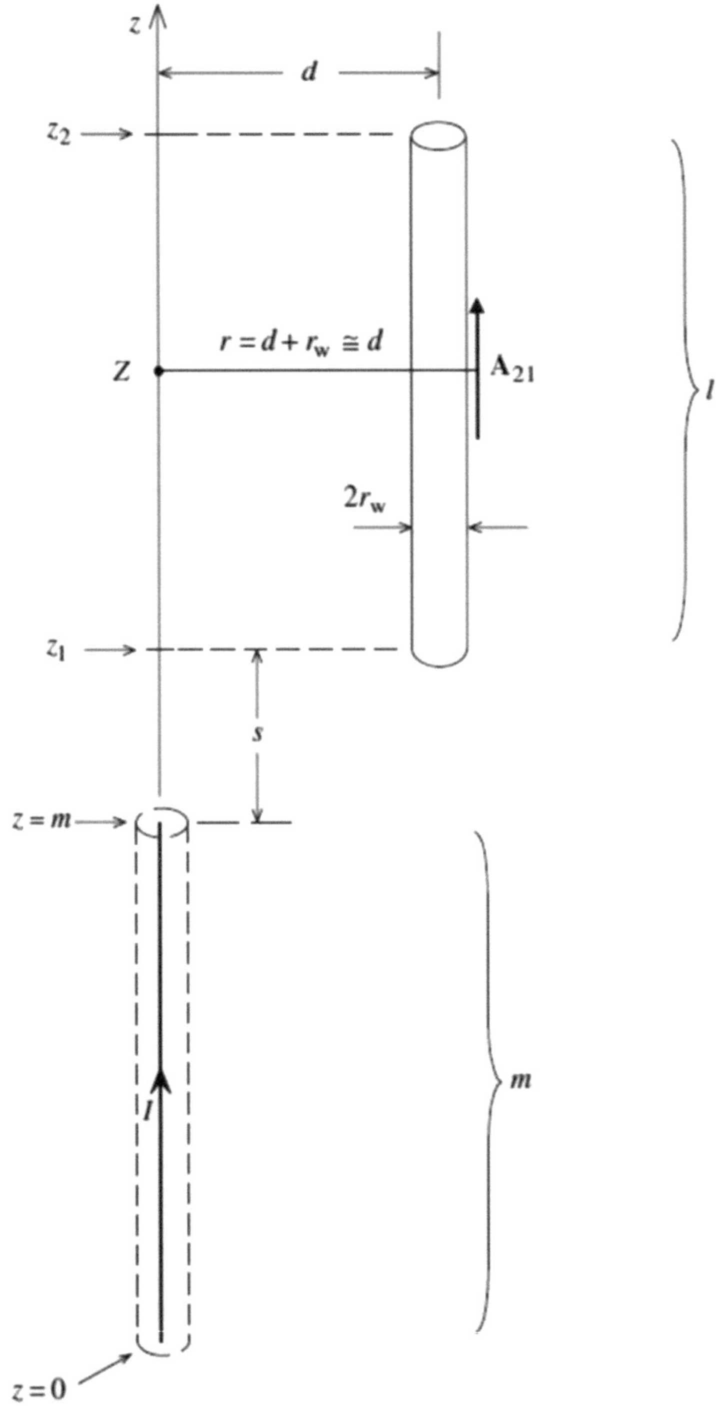

Figure 11.
Physical system for mutual inductance calculation.

2.1.3 Partial mutual inductance of two parallel offset wires

The calculation of the mutual inductance produced by two parallel wires like those ones shown in **Figure 11** is reported as per Eq. (19) [24]:

$$\tilde{M}_{\Phi} = \frac{\mu_0}{4\pi}\begin{bmatrix} z_2 \sinh^{-1}\left(\frac{z_2}{d}\right) - z_1 \sinh^{-1}\left(\frac{z_1}{d}\right) - (z_2 - m)\sinh^{-1}\left(\frac{z_2 - m}{d}\right) \\ + (z_1 - m)\sinh^{-1}\left(\frac{z_1 - m}{d}\right) - \sqrt{z_2^2 + d^2} + \sqrt{z_1^2 + d^2} \\ + \sqrt{(z_2 - m)^2 + d^2} - \sqrt{(z_1 - m)^2 + d^2} \end{bmatrix} \quad (19)$$

In **Figure 12**, the graphical representation of the inductance of two parallel wires located axially at a certain distance by s and orthogonally by d is shown. In the same figure, it is possible to see that in general the inductance decreases by increasing the geometrical distance since it depends on the magnetic field flow. However, the curves describing the inductance as a function of the distance d exhibit a decreasing rate that is higher by decreasing the axial distance between the wires which is more than the decreasing rate of the inductance as function of the axial distance by varying the axial distance along the orthogonal direction to the wire itself.

The combination of square roots and inverse hyperbolic sine function produces a double concave surface asymptotically tending to the infinity:

2.1.4 Mutual inductance of wires run by current and lying on the same line

The system shown in this paragraph is shown in **Figure 13** where both wires of finite length crossed by a generic amount lay coaxially:

The mutual inductance generated by such a configuration is described by Eq. (20) [24]:

$$M_{\Phi} = \frac{\mu_0}{4\pi}\begin{bmatrix} (l+s+m)\sinh^{-1}\frac{l+s+m}{r_w} - (m+s)\sinh^{-1}\frac{m+s}{r_w} - \\ +(l+s)\sinh^{-1}\frac{l+s}{r_w} + s\sinh^{-1}\frac{s}{r_w} - \sqrt{(l+s+m)^2 + r_w^2} \end{bmatrix} \quad (20)$$

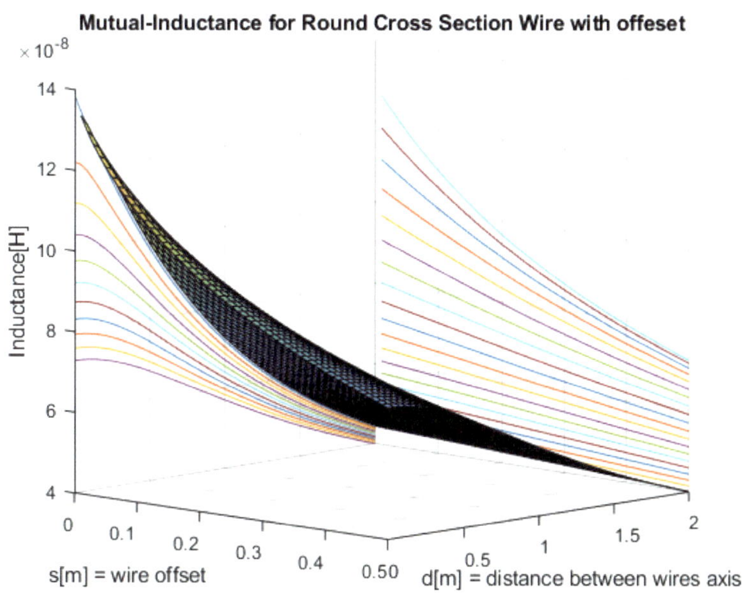

Figure 12.
Partial mutual inductance for two parallel wires.

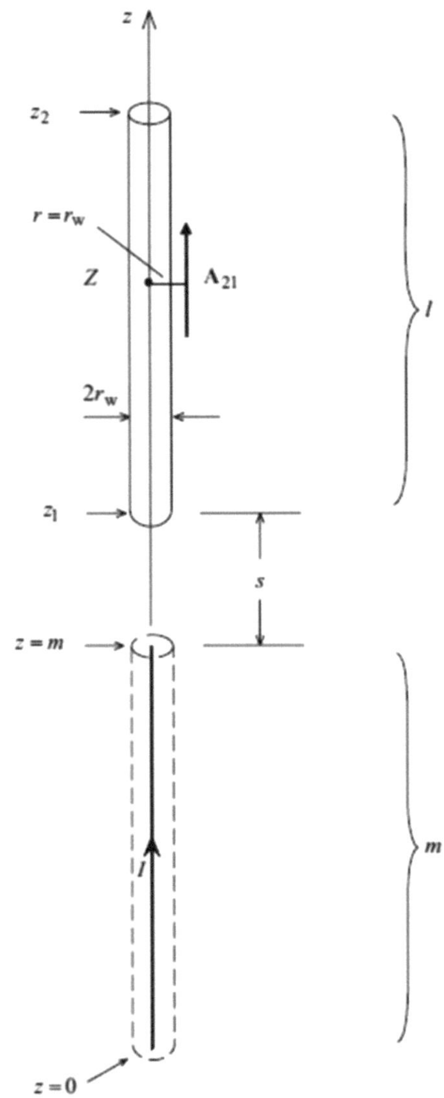

Figure 13.
Physical system for mutual inductance calculation.

that has been shown in **Figure 14** where the inductance as a function of the axial distance between the edges of two wires has been reported:

In **Figure 14** the inductance exhibits a decreasing of the inductance by increasing the distance *s* between two coaxial wires. This is the same concept that has been shown in the previous figures as demonstration that in case of the presence of current in a specific conductor pattern, there will always be an influence between those patterns coupled to a decreasing of inductance by increasing the distance between two conductors. These patterns can be considered as partial, characterized by an insulated wire for a specific length, or they can be coils constituting a closed loop. In this last case, it is necessary to consider the single contribution coming from each side of the closed loop in terms of auto-inductance and mutual inductance.

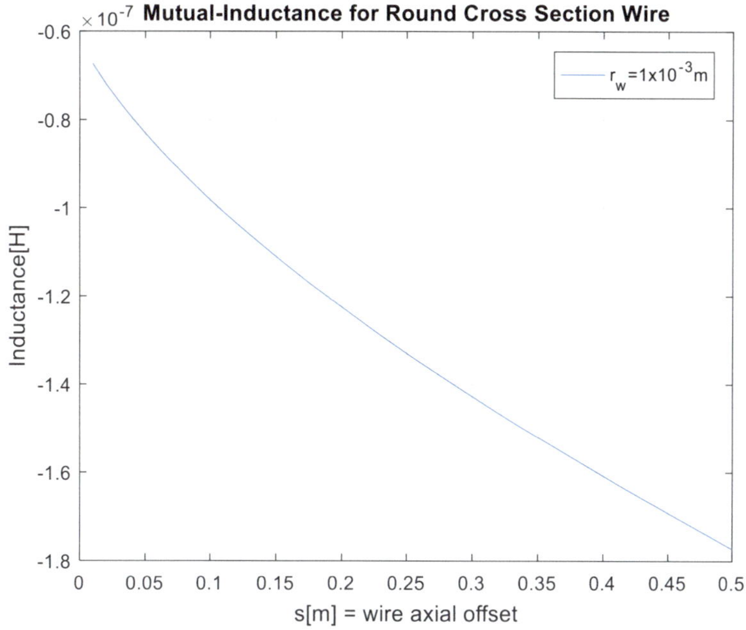

Figure 14.
Partial mutual inductance for two parallel wires.

3. Receiver coils

The receiving coil links the signal coming from the magnetization vector that rotates in space. This rotation generates an electromotive force that in the presence of an electrical resistance produces the electrical current. The current represents a signal characterized by a certain frequency content that processed through the discrete Fourier transform allows the images' generation. The circuit elements must also be able to store the electrical energy produced by a signal that in general can be decomposable in sinusoidal functions. For this reason, the receiving coils are designed in different configurations according to the frequency band of interest. It is for this founded reason the receiving coils are designed in three different possible ways: low-pass coils, high-pass coils, or band-pass coils. Two of the most applied

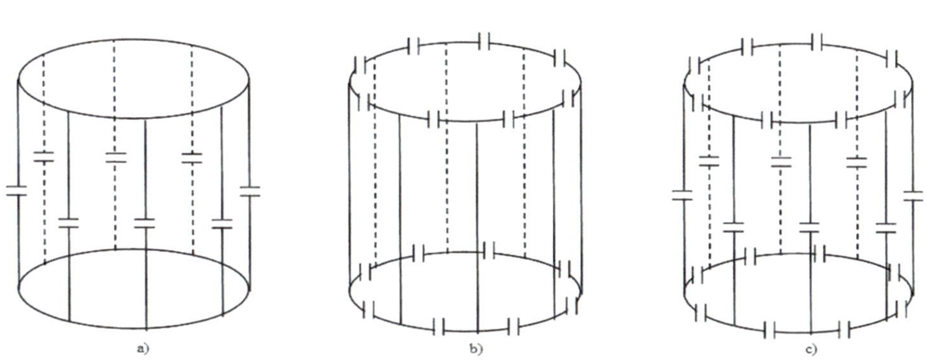

Figure 15.
Birdcage coil. (a) Low-pass, (b) high-pass, (c) band-pass.

configurations are the so-called birdcage coil and the phased array coils as shown, respectively, in **Figures 15** and **16**.

In **Figure 15** the coils are characterized by two rings joined by columns and rings, and columns are made of copper material. According to the desired frequency, the capacitors are in different positions of coils as shown in **Figure 15**. **Figure 16** shows the phase array configuration which is the Cartesian representation of birdcage coils with a small difference, or rather, there is an overlap of the circuits to reduce the electromagnetic coupling between the nearest coils [25], but for what concerns the filtering capabilities, the capacitors' placement reflects the birdcage coil according to the high-pass, low-pass, and pass-band filtering. The phased array structure is what will be considered to develop the experimental validation of results. In the next section, we will present first the analysis of a single loop as a derivation of the study developed in this section.

3.1 Single loop coil

In this section we determine the inductance of the rectangular loop shown in **Figure 17**, whose length is l and width is w. The conductors of the loop are rectangular flat having a width $2r_w$. We assume that the current I is uniformly distributed across the cross section of the wires, so that with regard to computing the magnetic field from it, the current can be considered to be concentrated in a filament on the axes of those wires. For isolated direct currents (dc) not in proximity to other

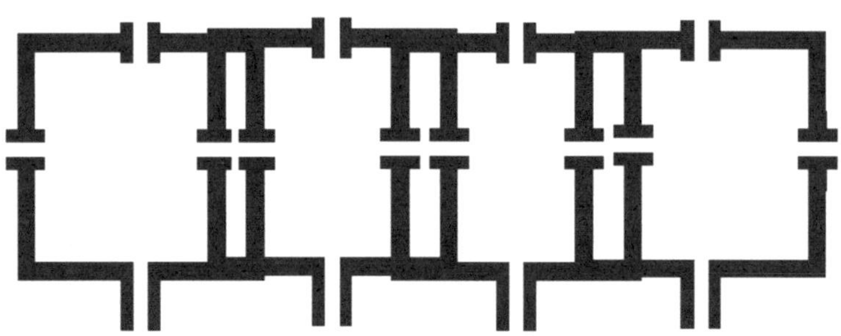

Figure 16.
Phased array coil: band-pass.

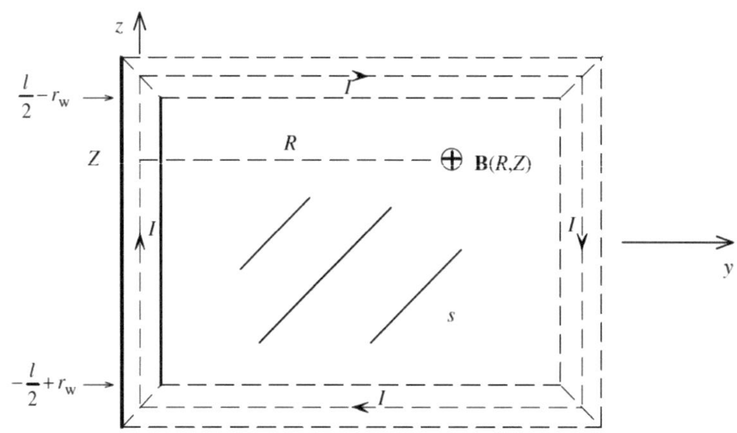

Figure 17.
Single loop coil.

currents, the current is, in fact, uniformly distributed over the wire cross section. However, for a current that is near other currents, the current in the wire will not be distributed uniformly over the wire cross section. Nearby currents will cause the current to be concentrated on the side of the nearest wire, a phenomenon known as the proximity effect. Proximity effect is usually not pronounced unless the two currents are within about four radii of each other (i.e., one wire will just fit between the two) [24]. To determine the total flux through that loop, we determine the flux through the loop caused by the current of each wire separately and then add the four fluxes:

$$L_{loop} = \frac{\sum_{i=1}^{4} \Phi_i}{I} = \frac{\sum_{i=1}^{4} \iint_i \vec{B}(y,z)dydz}{I} \tag{21}$$

where $\vec{B}(y,z)$ is the magnetic field expression Eq. (7) that is Eq. (22) for the vertical wire as per **Figure 16**:

$$\vec{B}(y,z) = \frac{\mu_0 I}{4\pi y}\left(\frac{z + l/2}{\sqrt{y^2 + (l/2 + z)^2}} - \frac{z - l/2}{\sqrt{y^2 + (l/2 - z)^2}}\right) u_\vartheta \tag{22}$$

The integral of vertical wires that is developed according to the integration edges as per Eqs. (23) and (24) is:

$$\Phi_v(l,w,r_w) = \Phi_1(l,w,r_w) + \Phi_3(l,w,r_w) = 2\int_{z=r_w-l/2}^{l/2-r_w}\int_{y=r_w}^{w-r_w} \vec{B}(y,z)dydz \tag{23}$$

$$\Phi_h(l,w,r_w) = \Phi_2(l,w,r_w) + \Phi_4(l,w,r_w) = 2\int_{z=r_w-w/2}^{w/2-r_w}\int_{y=r_w}^{l-r_w} \vec{B}(y,z)dydz \tag{24}$$

where $\Phi_1(l,w,r_w) = \Phi_3(l,w,r_w)$ and $\Phi_2(l,w,r_w) = \Phi_4(l,w,r_w)$.

According to the integral table [26] and dividing the total magnetic flow for the current I, the inductance for the entire loop is carried out as:

$$\begin{aligned} L(l,w,r_w)_{loop} &= \frac{\Phi_v(l,w,r_w) + \Phi_h(l,w,r_w)}{I} \\ &= \frac{\mu_0}{\pi}\left[\begin{array}{l} (r_w - l)\sinh^{-1}\frac{l - r_w}{w - r_w} + (r_w - w)\sinh^{-1}\frac{w - r_w}{l - r_w} \\ + (l - r_w)\sinh^{-1}\frac{l - r_w}{r_w} + (w - r_w)\sinh^{-1}\frac{w - r_w}{r_w} \\ + r_w\sinh^{-1}\frac{r_w}{w - r_w} + r_w\sinh^{-1}\frac{r_w}{l - r_w} \\ + 2\sqrt{(l - r_w)^2 + (w - r_w)^2} - 2\sqrt{(w - r_w)^2 + (r_w)^2} \\ - 2\sqrt{(l - r_w)^2 + (r_w)^2} - 2r_w\ln\left(1 + \sqrt{2}\right) + 2\sqrt{2}r_w \end{array}\right] \end{aligned} \tag{25}$$

In **Figure 18** the inductance produced by a closed loop varying its dimension has been shown.

In the same figure, the inductance increases by increasing the area enclosed by the loop. So, the higher is the surface area, the higher is the induction phenomena the magnetization vector generates. An important information is carried out by the

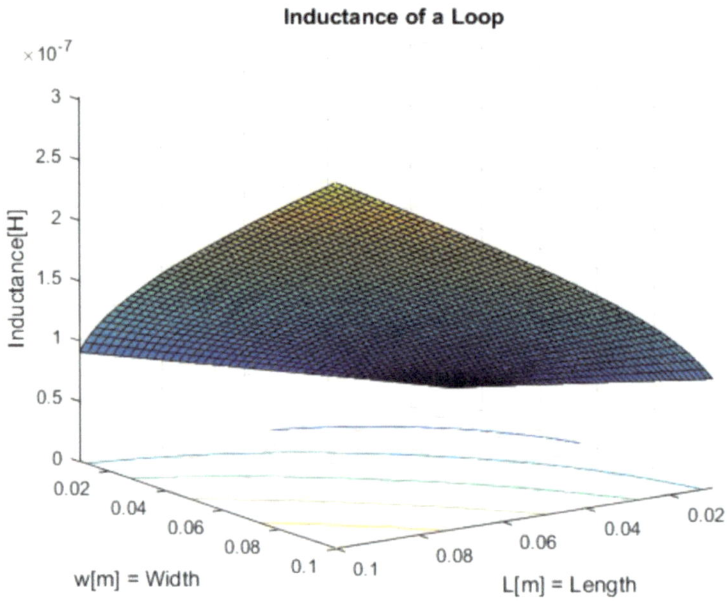

Figure 18.
Inductance of a rectangular lop varying the side dimension.

frequency response (FR) characterizing the loop that can be considered as an LC circuit having the voltage expression as in Eq. (26):

$$V(\omega) = \left[j\omega L(l, w, r_w)_{loop} - \frac{j}{\omega C_x}\right] I(\omega) = j\left[\frac{\omega^2 L(l, w, r_w)_{loop} C_x - 1}{\omega C_x}\right] I(\omega) \quad (26)$$

In Eq. (25) the capacitor is the variable that allows the shifting of peak of resonance (*pr*) in the FR diagram.

3.1.1 Frequency tuning and magnitude variation

The fact that we are introducing a single loop means there is only one frequency of resonance once the capacitor value is chosen. The value of the resonance $\omega_{res} = 1/\sqrt{L(l, w, r_w)_{loop} C_x}$ that will produce the singularity in the magnitude diagram in a bode plot according to the transfer function (*tf*) is described in Eq. (27):

$$\left|\frac{I(\omega)}{V(\omega)}\right|_{dB} = 20Log\left|\frac{\omega C_x}{\omega^2 L(l, w, r_w)_{loop} C_x - 1}\right| \quad (27)$$

The FR is carried out as shown in **Figure 18** for a loop having a square shape with $l = w = 10^{-1}$m and $r_w = 3.5 \times 10^{-3}$m varying the capacity in a certain value and neglecting the resistance effect. In **Figure 18** it can be seen that an increasing of capacitive value corresponds to a decreasing of frequency of resonance (*fr*) value and an increasing of the magnitude at that frequency. The range of frequency where the *Bode* plot has been shown goes from 10 MHz to 10GHz with a useful signal amplification between 50 MHz and 120 MHz. According to the same figure, the frequency of resonance is always the frontier where on its right the curve is decreasing, while on the left the answer increases. It can be seen that by

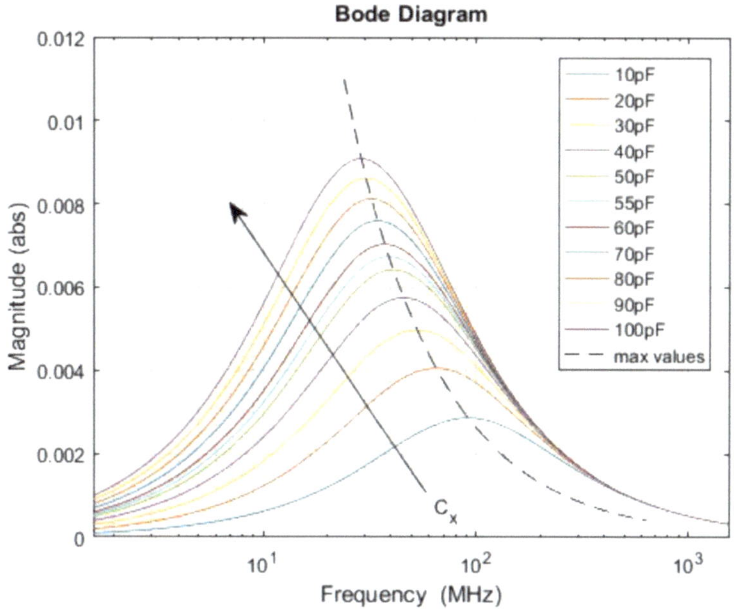

Figure 19.
Frequency response of a single coil with variable capacitor.

differentiating the *tf* in Eq. (27), the condition of these two regions separated by the value $\omega_{res} = 1/\sqrt{L(l, w, r_w)_{loop} C_x}$ is carried out. Eq. (27) establishes the effect of the capacitor is double, or rather, it varies the resonance frequency; it moves up and down the *pr*, decreasing the power of the receiving signal at high frequency.

In **Figure 19**, we can consider the FR referred to a capacitive value of $C_0 = 55\text{pF}$ in order to perform some analysis that allows to understand better the dynamic of a single receiver coil. A desired value of *fr* called ω_r that is found to be at the right of the *fr* corresponding to the capacitive C_0 that we call $\omega_0 = 40\text{MHz}$ satisfies the condition that $\omega_r > \omega_0$. In this case we write the system with $C_0 = 1/L_{loop}\omega_0^2$, $C_r = 1/L_{loop}\omega_r^2$, and $C_r = C_0 + \Delta C_{0r}$ carrying out the relation as:

$$\Delta C_{0r} = \frac{1}{L_{loop}} \left(\frac{\omega_0^2 - \omega_r^2}{\omega_0^2 \omega_r^2} \right) \tag{28}$$

Eq. (27) establishes that in order to reach the desired frequency $\omega_r > \omega_0$, the capacitive value has to decrease since $\Delta C_{0r} < 0$. A decreasing of capacitor value implies a decreasing of the magnitude as per Eq. (27); this effect can be seen also by considering the curve describing the *pr* varying the frequency in Eq. (29) obtained after a differentiation of the *tf* and substituting the capacitive value as a function of frequency in Eq. (27) and shown in a dashed line joining all the *pr* in **Figure 18**:

$$\hat{W}(\omega) = \frac{1}{2L_{loop}\omega} \tag{29}$$

The manipulation of Eq. (29) with the same methodology that led to Eq. (28) returns the expression shown in Eq. (30):

$$\Delta \hat{W}_{or} = \frac{1}{2L_{loop}} \left(\frac{\omega_0 - \omega_r}{\omega_0 \omega_r} \right) < 0 \tag{30}$$

Eq. (30) establishes that for $\omega_r > \omega_0$ the magnitude of the FR decreases or rather $\hat{W}_r(\omega_r) < \hat{W}_0(\omega_0)$.

The case where $\omega_r < \omega_0$ produces opposite sign in the differences ΔC_{0r} and $\Delta \hat{W}_{or}$, but the relation between the ΔC_{0r} and $\Delta \hat{W}_{or}$ remains the same as in Eq. (31):

$$\frac{\Delta \hat{W}_{or}}{\Delta C_{0r}} = \frac{1}{2} \frac{\omega_0 \omega_r}{\omega_0 + \omega_r} \left[\frac{\mathrm{dB}}{\mathrm{F}}\right] \tag{31}$$

One has the infinitesimal differentiation for $\Delta C_{0r} \to 0$ from where it has $\omega_r \cong \omega_0$

$$\frac{d\hat{W}_{or}}{dC_{0r}} = \frac{1}{4}\omega_r \left[\frac{\mathrm{dB}}{\mathrm{F}}\right] \tag{32}$$

The receiver coil behaves as a low-pass filter since it tends to amplify the signal having a low frequency in a certain operating frequency established by the possible values that the capacitor might have.

4. Conclusions

In this chapter, a meticulous study on the basic design of a receiver coil characterized by a single turning coil has been developed. The study has started by introducing the technique used in calculating the inductance produced by different shapes of wire crossed by a constant value of current. The inductance is a fundamental parameter that affects the performance in static and dynamic conditions, revealing how the intensity of magnetic field produced by the coil may interfere, through the mutual inductance to the other branches of the same coil. The frequency response of a given single coil loop corresponding to a value of capacitor is characterized by a single peak of resonance that shifts varying the capacitor tuned for a nominal frequency ω_0. The frequency resonance shifts are due to the capacitive effect and have as second consequence a variation of the peak of resonance. The variation of resonance frequency is a desired effect the author wants to reach in order to capture a signal having that frequency, while a numerical strategy has been introduced in order to quantify the variation of the magnitude that will affect the output signal produced by the coil.

Author details

Gabriele Barbaraci
University of Palermo, Palermo, Italy

*Address all correspondence to: gabriele.barbaraci@yorku.ca

References

[1] Stumpf C, Malzacher M, Schmidt LP. Radio frequency modeling of receive coil arrays for magnetic resonance imaging. Journal of Imaging. 2018;**4**(5): 67. DOI: 10.3390/jimaging4050067

[2] Barbaraci G, D'Ippolito F. An estimator algorithm for the rotation time of magnetization vector in nuclear magnetic resonance for imaging (NMRI). Journal of Vibration and Control. 2018:1-22. DOI: 10.1177/1077546318791608

[3] Jin J. Electromagnetic Analysis and Design in Magnetic Resonance Imaging. Florida: CRC Press; 1999. ISBN 9780849396939- CAT# 9693

[4] Idziak S, Haeberlen U. Design and construction of a high homogeneity rf coil for solid-state multiple-pulse NMR. Journal of Magnetic Resonance. 1982;**50**: 281-288

[5] Ibrahim TS, Lee R, Baertlein BA, et al. Effect of RF coil excitation on field inhomogeneity at ultra high fields: A field optimized TEM resonator. Magnetic Resonance Imaging. 2001;**19**: 1339-1347

[6] Peck TL, Magin RL, Lauterbur PC. Design and analysis of microcoils for NMR microscopy. Journal of Magnetic Resonance, Series B. 1995;**108**: 114-124

[7] Webb AG. Radiofrequency microcoils for magnetic resonance imaging and spectroscopy. Journal of Magnetic Resonance. 2013;**229**:55-66. DOI: 10.1016/j.jmr.2012.10.004

[8] Kc R, Henry ID, Park GHJ, et al. Design and construction of a versatile dual volume heteronuclear double, resonance microcoil NMR probe. Journal of Magnetic Resonance. 2009; **197**:186-192. DOI: 10.1016/j.jmr.2008.12.020

[9] Watzlaw J, Glöggler S, Blümich B, et al. Stacked planar micro coils for single-sided NMR applications. Journal of Magnetic Resonance. 2013;**230**: 176-185. DOI: 10.1016/j.jmr.2013.02.013

[10] Kozlov M, Turner R. Fast MRI coil analysis based on 3-D electromagnetic and RF circuit co-simulation. Journal of Magnetic Resonance. 2009;**200**:147-152. DOI: 10.1016/j.jmr.2009.06.005

[11] Bart J, Janssen JWG, van Bentum PJM, et al. Optimization of stripline-based microfluidic chips for high-resolution NMR. Journal of Magnetic Resonance. 2009;**201**:175-185. DOI: 10.1016/j.jmr.2009.09.007

[12] Qian C, Masad IS, Rosenberg JT, et al. A volume birdcage coil with an adjustable sliding tuner ring for neuroimaging in high field vertical magnets: Ex and in vivo applications at 21.1 T. Journal of Magnetic Resonance. 2012;**221**:110-116. DOI: 10.1016/j.jmr.2012.05.016

[13] Georget E, Luong M, Vignaud A, et al. Stacked magnetic resonators for MRI RF coils decoupling. Journal of Magnetic Resonance. 2017;**275**:11-18. DOI: 10.1016/j.jmr.2016.11.012

[14] Ohliger MA, Sodickson DK. An introduction to coil array design for parallel MRI. NMR in Biomedicine. 2006;**19**:300-315. DOI: 10.1002/nbm.1046

[15] Rogovich A, Monorchio A, Nepa P, et al. Design of magnetic resonance imaging (MRI) RF coils by using the method of moments. In: IEEE Antennas and Propagation Society Symposium. 2004. DOI: 10.1109/APS.2004.1329829

[16] Stara R, Fontana N, Alecci M, et al. RF coil design for low and high field MRI: numerical methods and measurements. In: IEEE Nuclear Science

Symposium Conference Record. 2011. DOI: 10.1109/NSSMIC.2011.6152634

[17] Xu B, Wei Q, Liu F. An inverse methodology for high-frequency RF coil design for MRI with de-emphasized B_1 fields. IEEE Transactions on Biomedical Engineering. 2005;**52**(9). DOI: 10.1109/TBME.2005.851514

[18] Abraham R, Ibrahim TS. Human-body coil design for magnetic resonance imaging at 7 Tesla. In: IEEE Antennas and Propagation Society International Symposium. 2005. DOI: 10.1109/APS.2005.1551696

[19] Li BK, Xu B, Hui HT, et al. A new approach for magnetic resonance RF head coil design. In: IEEE Engineering in Medicine and Biology 27th Annual Conference. 2005. DOI: 10.1109/IEMBS.2005.1615624

[20] Morey AM, Bhujade S, Bhuiya T. Design and development of surface coil for 1.5T MRI scanner. In: International Conference on Smart Technologies and Management for Computing, Communication, Controls, Energy and Materials (ICSTM). 2015. DOI: 10.1109/ICSTM.2015.7225428

[21] Shou G, Xia L, Liu F, et al. MRI coil design using boundary-element method with regularization technique: A numerical calculation study. IEEE Transactions on Magnetics. 2010;**46**(4)

[22] Basari ASH, Rahardjo ET, Zulkifli FY. Eight-channel phased array RF coils design for 3T parallel MRI system. In: IEEE 4th Asia-Pacific Conference on Antennas and Propagation (APCAP). 2015. DOI: 10.1109/APCAP.2015.7374269

[23] Wei S, Yang W. A high-temperature superconducting RF coil design for low field MRI. In: Asia-Pacific International Symposium on Electromagnetic Compatibility (APEMC). 2016. DOI: 10.1109/APEMC.2016.7522954

[24] Paul CR. Inductance-Loop and Partial. John Wiley & Sons, Inc.; 2010. ISBN: 978-0-470-46188-4

[25] Brown RW, Cheng YN, Haacke EM, et al. Magnetic Resonance Imaging: Physical Principles and Sequence Design. John-Wiley & Sons, Inc.; 2014

[26] Dwight HB. Tables of Integrals and Other Mathematical Data. 4th ed. New York: Macmillan; 1961

Chapter 3

Facile NMR Relaxation Sensor for Monitoring of Biomass Degradation Products during Conversion to Biogas

Wiesman Zeev and Linder Charles

Abstract

The chemical and morphological composition of animal biowaste is known to limit the efficiency of methane production by bacterial anaerobic digestion (AD). To better understand these material limitations, we studied degradative changes in cattle manure's organic complex components chemical and morphological composition during its AD to methane. This was achieved using low field ^{1}H NMR relaxation times domain (TD) spectral mapping combined with T_1 (spin-lattice) and T_2 (spin-spin) TD of cattle manure biomass (CM) peaks assignment, starting from samples of initial freshly collected CM biomass sample followed by several time points sampling during 21 days cycle of the AD process. A T_1-T_2 relaxation TD graph giving a stable reproducible pattern of 12 peaks was generated, and assigned to different domains, whose changes during AD could be observed. These 12 peaks were assigned to TDs of crystalline nano-aggregated complexes of different degrees of crystallinity with low porosity and low hydration rate and a morphological group of amorphous domains with increased pore size, density, and higher hydration. In agreement with models of elementary cellulose fibrils, these domains were designated as three layers of cellulose consisting of interior, subsurface, and surface. The most amorphous TD volume showed good correlation with biogas production and could serve as an indicator for digestibility and cellulose conversion to a glucose intermediate during the AD process. This study demonstrated the facile and versatile usage of 2D ^{1}H NMR T_1-T_2 sensorial technology in studying complex biowaste systems, with the potential for improving CM biomass conversion efficiency into bio-methane.

Keywords: anaerobic digestion (AD), biogas, lignocellulose, ^{1}H LF-NMR relaxation, time domain (TD), cellulose crystallinity, biofilm

1. Introduction

To meet the needs of a growing world population, fossil fuels limited supply and global warming by greenhouse gas emissions, economically competitive biofuels with neutral greenhouse effects are being developed such as bacterial anaerobic digestion (AD) of biomass into methane [1, 2]. Although extensive studies have been carried out to maximize AD methane production, the major limitation is still the relatively low yield and biomass conversion rate [3]. During AD, the bacterial enzymes hydrolyze large organic molecules into derivatives, which are then metabolized into mainly methane (55–70%) and carbon dioxide (45–30%) [1, 3]. Bacterial growth forms biofilms of colonies enclosed by extracellular polymeric substances (EPS), which films are believed to have a positive effect on bacterial activity and methane production in AD systems. However, its exact mechanism is not well known [4–7].

Cattle manure (CM) biowaste is one of the most common available feedstock for AD, but its low yield and conversion efficiency into bio-methane significantly reduces its cost effectiveness as a renewable fuel source [2, 4]. In CM, there are relatively high concentrations of cellulosic material compared to the plant forage, due to removal during forage digestion of free sugars and fats. CM contains about 40% cellulose and hemicellulose with smaller amounts of lignin, proteins, and lipids [8]. This cellulosic material is relatively resistant to bacterial AD hydrolysis [9]. This unique structural arrangement of plant cell wall cellulose and its morphological changes during AD greatly affects the hydrolysis and degradation of CM to produce biogas. Plant cell wall cellulose, contains highly ordered cellulose microfibrils bound by both inter and intramolecular forces. These cellulose microfibrils assemble into fibers with highly ordered crystalline domains, known as Iα and Iβ, interspersed within amorphous matrix regions of hemicellulose (polysaccharide) and lignin (polyphenol)—together called lignocellulose [10, 11].

Surprisingly, while there are extensive parametric studies of hydrolysis (AD and fermentation) of cellulosic biomass such as cotton and sugar cane to produce fuel such as methanol and methane [11], there is only limited research on AD of cellulosic CM. Mechanism and structural studies of fuel generation by conversion of cellulosic biomass with high cellulosic content such as cotton and sugar cane have used well-known analytical methods such as Fourier-transform infrared spectroscopy (FTIR), thermogravimetric analysis (TGA), X-ray diffraction (XRD), scanning electrons microscope (SEM), and high-resolution ^1H and ^{13}C nuclear magnetic resonance (NMR) [12]. These conventional methods are limited in on-line monitoring of molecular structure identification during pretreatment (e.g., acid or base hydrolysis, [13, 14], of the starting materials) and during AD. These analytical methods can either measure the crystallinity index or chemical group identification to correlate material characteristics to fuel production efficiency, i.e., methanol and methane. However, enzyme accessibility to given sites is affected not only by crystallinity and ratio of cellulose to lignin but also by specific morphological domain size, its surface structure, porosity, and their changes during hydrolysis [15, 16].

Proton low-field energy relaxation NMR (^1H LF-NMR) is a nondestructive method for detecting molecular and morphological changes of aggregates and complexes by two different proton relaxation times T_1 and T_2. T_1 is a longitudinal relaxation time affected mainly by the interaction between spins and surrounding environment (spin–lattice); T_2 is a transverse relaxation time affected by reactions between neighboring spins (spin–spin). T_1 and T_2 measure energy exchange within the molecular environment related to interaction energies. These values can be utilized to identify different morphological aggregate assemblies [17, 18]. Water protons (^1H) are readily measured by ^1H LF-NMR and in different environments have different energy relaxation times that can be used to identify different morphologies [19]. Within free water and liquid bulk T_1 equals T_2, therefore, measurement of differences between T_1 and T_2 can characterize liquid-solids interactions. Since T_2 relaxation time is more strongly affected than T_1, these differences are used to measure the water-surface interactions [20–24]. Three water phases are conventionally described in solids: bound water, gel water, and capillary water. Bound water are non-freezing water chemically attached to the surface and exhibit extremely low T_2. Water in the gel pores is more mobile, but the relaxation is heavily dominated by surface interactions, giving a low T_2. The T_2 of ^{1}H of water in larger capillary pores is longer, but still limited and significantly lower than the bulk liquid relaxation times. The T_1 relaxation time is governed mainly by proton-surface interactions, but it is also affected by the water phase, which may be correlated to different morphologies and crystalline/amorphous ratios [23, 25].

The ratio T_1/T_2 has proven to be a valuable tool for measuring proton-molecular substrate interactions without being sensitive to material parameters such as pore density, surface area, and paramagnetic particles, which would complicate the analysis [20, 21]. T_1/T_2 can be used to quantify how morphologies affect the interaction strength and their changes under different conditions [20]. For example, this ratio was identified as a parameter of molecular substrate interactions intensity, and a measure of the activation energy of adsorption of solvents on different substrates and porous structures [21].

In this paper, we address the currently low AD conversion of cellulosic components of CM biowaste to biogas by mapping and quantification of the CM lignocellulosic nano-aggregated internal structures, at different times of AD using a novel 2D ^{1}H LF-NMR T_1-T_2 proton energy relaxation time-domain analysis recently developed in our laboratory. This approach to material analysis was first described in Wiesman et al. [19] for CM, wherein ^{1}H LF-NMR energy relaxation times T_1 and T_2 and T_1/T_2 ratios are used to form 2D and 3D chemical and morphological graphs and a corresponding T_1-T_2 TD peak assignment to material composition. In the present study, 2D ^{1}H LF-NMR T_1-T_2 energy relaxation mapping and T_1/T_2 ratios of absorbed water ^{1}H are used to identify nano-chemical and morphological structures in the CM biomass starting from cattle forage (CF) as a non-digested control, through to digested CM products and subsequent AD samples of CM under industrial mesophilic AD pilot conditions.

2. Experimental

2.1 Anaerobic digestion system

Cattle manure was collected from a district plant treating 600 tons of CM per day, from the underground preliminary tank, where the raw cattle manure is mixed with water to total solids of ~10%. AD experiments were performed in a small scale pilot plant in a laboratory at I.T.B. Ltd. Israel. The pilot plant consists of 30 L digester with temperature control, mixers, and biogas flow meters. All digestion tests were performed at batch and mesophilic conditions for a period of 21 days. Samples were taken for chemical and physical analysis before AD (time 0) and after 3, 6, and 24 h and 8 and 21 days of AD. Solids analysis of total solids (TS) and volatile solids (VS) were carried out according to standards methods [1–3]. Biogas flow was measured with Aalborg Mass Flow Meter GFM17. The flow-meter values were collected and saved by the software Real Time Viewer 1.43 Brain Child Electronics Co., Ltd.

Forage samples were collected from a feedlot of a dairy cowshed. The forage is composed of about 60% hay, 30% silage (fermented grasses and grains), and 10% grains.

2.2 ^{1}H low field NMR (^{1}H LF-NMR)

2.2.1 Sample preparation

All samples were completely dried to constant weight at 65°C. Then ground to 0.1–1 mm particle size and kept in a closed glass jars. Dried and grinded bio-waste samples were wetted to 20% moisture content (mg-water/mg-total) and placed in an NMR glass tube for conditioning at 40°C on a heating plate for 90 min prior to the ^{1}H LF-NMR test.

2.2.2 Signal acquisition

^{1}H LF-NMR measurements were carried out on a Maran bench-top pulsed NMR analyzer (Resonance Instruments, Witney, UK) equipped with a permanent magnet and a 18-mm probe head operating at 23.4 MHz. One dimensional (1D) T_1 relaxometry experiments were performed by repeating an inversion recovery step [180°– t_1] for a determined number of runs, where T_1 is varied logarithmically between runs. The t_1 is a NMR signal acquisition time [26]. The T_1 period is dominated by spin–lattice longitudinal relaxation, including possible longitudinal cross-relaxation processes. T_2 relaxometry experiments were performed using a Carr-Purcell-Meiboom-Gill sequence (CPMG) pulse sequence. This multiple sequence consists of applying a single 90° pulse followed by multiple consecutive 180° pulses. This allows measuring transverse relaxation, which results from spin–spin interactions.

The 2D cross-correlation experiments were performed by a T_1-T_2 sequence signal analysis, where the inversion recovery step [180°– t_1] is inserted prior to the CPMG sequence [20]. As in the 1D analysis for 2D, the energy relaxation processes are a T_1 period dominated by matrix longitudinal relaxation mechanisms with a possible cross-relaxation processes; while the T_2 by spin–spin transverse relaxation mechanism [17].

2.3 Computing method for analysis of 2D T_2-T_1 energy relaxation TD

The computing methodology of a reconstruction algorithm for qualitative (peak location) analysis of 2D and 3D T_1-T_2 energy relaxation graphs used in the present study is fully described in previous publications [19, 27, 28].

For quantitative analysis of the 3D T_1-T_2 vs. peak volume, we used semi-automatic software procedures for separating each peak in the graph, as described in detail in previous report [19].

2.4 Peak assignment

Peak assignment was carried out by two following approaches: spiking with standards of lignocellulose components and chemical removal of separate lignocellulosic components. Spiking of peaks was made by addition of the following standards purchased from Sigma Aldrich: cellulose fibers medium, xylan (A hemicellulosic cell wall polysaccharide, lignin, pectin.) Additional materials used in this study included: pure standards of fatty acids (acetic, propionic, butyric, and valeric acid) and internal laboratory made nano-aggregated epoxy-polymer complexes prepared from oxidized polyunsaturated fatty acid, that was previously characterized [29] and shown to consist of cross-linked aldehyde and epoxy groups, was also used as a standard. The standards were added to the dried ground samples prior to re-watering, and then water was added to the mixture to the desired moisture content. The addition of chemical standards to CM was reflected in spiking of specific peaks/areas in the ^1H LF-NMR graphs, which enabled characterization of the lignocellulosic T_2-T_1 graphic peak sites in the CM [19].

Peak identification by modification of signal peaks of ^1H LF-NMR was made by chemically extracting/removal of lignocellulosic components according the standard approved protocols as follow: Diluted acid (HCl 2%) treatment was used for hemicellulose hydrolysis; lignin chemical degradation was based on H_2SO_4 72% according to Kalson method (ASTM D-1106) and Kurschner methodology with HNO_3 70% was used for Cellulose determination [30].

2.5 Peak volume quantification of ^1H

Quantification analysis of peak volume was made by testing the samples in ^1H LF-NMR with a fixed receiver gain (RG). RG controls the amplifying of signals acquired by the NMR. Therefore, the RG must be kept constant in order to quantify NMR signals with the same amount of amplification. All tests of cattle manure used in quantification of NMR signals were performed at RG = 3. A calibration curve of cellulose was performed by testing samples with different amount of a cellulose standard (Sigma-Aldrich C6288) with constant RG = 3.

2.6 Scanning electrons microscope (SEM)

The samples were spread onto conductive carbon tapes fixed on aluminum stubs and were sputtered with gold using Polaron Equipment Limited SEM Coating Unit E5100. The SEM micrographs were acquired in a Quanta 200 FEI Company SEM, with backscattered electron SEM imaging, secondary electron SEM imaging and energy dispersive X-ray spectroscopy (EDS) analysis.

2.7 Statistical analysis

The accuracy of 2D T_1-T_2 analysis was carried out in three repeated measurements for each of the six times (T-0, 3, 6, 24, 192, and 504 h) and analyzed with respect to the distribution of T_1 values, T_2 values, and the relative volume of each of the selected 10 peaks, i.e., peaks 1, 2, 3, 4, 5, 6, 7, 9, 10, and 11 (peak 8 was not detected in some tests and therefore omitted). We have analyzed the data of the six different times, getting a total of 18 observations for each parameter (T_1, T_2, relative volume) and a standard deviation of a statistical estimator was used to evaluate the distribution of each parameter, as described in detail in a previous paper [19].

3. Results and discussion

3.1 Characterization of CM conversion into biomethane in a mesophilic anaerobic reactor

To demonstrate the advantages of T_1-T_2 graphs generated by ^1H LF-NMR, for understanding AD mechanisms, we initially characterized the CM anaerobic digestion system using conventional material weight of volatile and nonvolatile solids. We measured total and volatile solids (TS and VS, respectively) in CM during 21 days of AD process (**Table 1**).

The initial amount of TS and VS loaded into the reactor was about 123 and 94 g/l, respectively. At the end of the AD process after 21 days, the amount of TS and VS was reduced by about 21 and 24% (wet weight), respectively.

Time	TS (g/l)	VS (g/l)
0 h	122.8	93.8
24 h	120.1	93.3
192 h	112.4	86.1
504 h	97.2	71.1
% Removal	20.8	24.2

Table 1.
Total and volatile solids during anaerobic digestion of cattle manure.

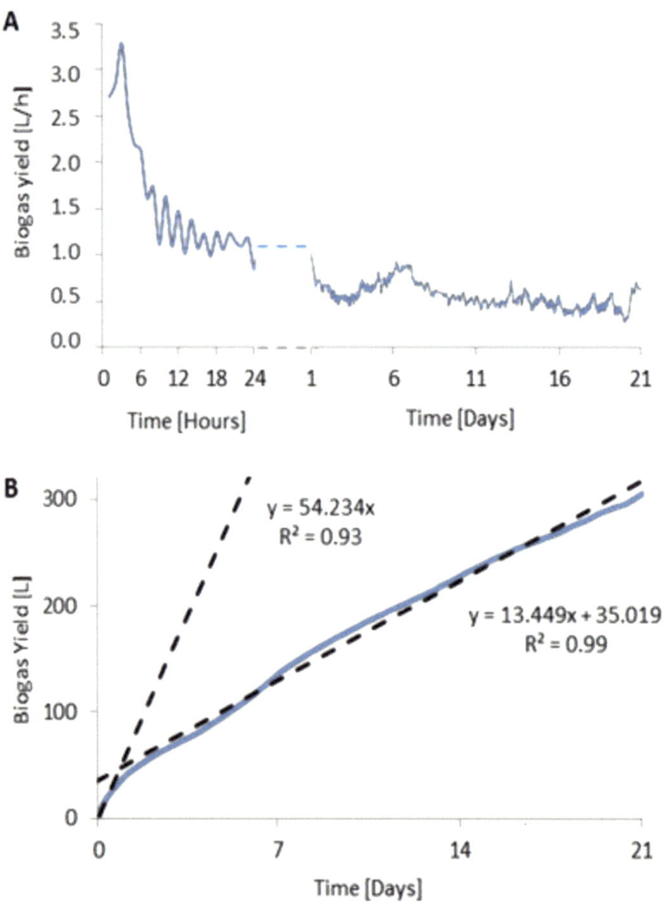

Figure 1.
Biogas yield during 21 days (504 h) of mesophilic batch AD of CM. (A) Daily biogas flow rate. The first 24 h of digestion are the most productive, in which the biogas flow rate reaches the maximum (CM-3 h) and then sharply decreases. (B) Cumulative biogas yield during 21 days of anaerobic digestion. The trend line slope of the first 24 h is much steeper than the slope of the later hours indicating the efficiency of the first 24 h.

Figure 1A shows hourly biogas flow with a typical peak seen in the first 3 h, which continually decays for 24 h. Additionally, biological enzymatic processes are often characterized with fluctuations, which are reflected in the biogas flow. **Figure 1B** shows cumulative biogas flow with a steep slope in the first 24 h and moderate slope for the later days. These biogas characteristic curves are in agreement with other studies [31]. Although the curve slope of cumulative biogas yield decreases and usually reaches a plateau after several weeks industrial AD continues beyond the optimal biogas production time, to reduce residual product waste pathogen, odors, and pollutants to environmental none restrictive levels [2, 32].

Overall, 21 days biogas yield was 306 or 155 L per kg-VS. The biogas composition was mainly methane (57.7%) and carbon dioxide (37.7%) with a small amount of H_2S (5000 ppm), in good agreement with the prior studies [31, 33]. Biogas efficiency is commonly measured by removal of VS, which represent the organic fraction of biomass; nevertheless biomass with the same initial VS may show different amounts of VS removed [31, 33].

It should be noted that the remaining ~80% of the non-converted CM biomass contains organics that are not accessible to degrading enzymes, because of morphological or chemical barriers, and therefore are more resistant to bacterial digestion and are disposed as soil amendment.

3.2 Qualitative monitoring of CM biomass degradation in mesophilic anaerobic reactor

In the present study using 2D T_1-T_2 energy relaxation chemical-morphological mapping technology, the goal was to better understand the parameter/variable factors resulting in the relatively low conversion rate of CM lignocellulose into methane, as shown above (**Figure 1A** and **B**), and to demonstrate a potentially cost effective accurate biosensor to monitor and control the AD process. The objective is to optimize the energy conversion potential of CM by increasing bio-methane yield, by reducing chemical and morphological barriers to the degrading enzymes by using the mapping, assignment and quantification of the different CM lignocellulose nano-aggregated structures and complexes initially and at different AD time periods.

3.2.1 T_1-T_2 mapping of CM biomass decomposition during 21 days of a mesophilic anaerobic reaction

As described in the introduction by adding water, it is possible to study the porosity of polymeric materials using ^1H LF-NMR to generate proton T_1 and T_2 energy relaxation time domains [19, 23, 34]. T_2 spin–spin energy relaxation of water's ^1H provides a good indication of pores-size distribution in lignocellulosic materials [35, 36]. T_1 spin–lattice energy relaxation constant characterizes the spin-matrix interactions of ^1H within the lignocellulose fraction. These interactions are rationalized by hydrogen bonds of water's protons and energy exchange with the OH groups of the cellulosic components [19, 36]. To generate reproducible T_2 and T_1 energy relaxation time domain using ^1H LF-NMR, the water concentration in the lignocellulosic material should be an optimum needed for uniform thin-surface coatings to increase the water's ^1H T_1 signal, corresponding to the interaction with the outer lignocellulose complexes matrix, while simultaneously not filling pore volume, for generating well resolved signals of the spin–spin (T_2) energy relaxation of surface adsorbed water molecules. Previous studies showed that adding water to lignocellulose samples up to 20% is optimal for T_1-T_2 energy relaxation characterization [12, 19, 35].

Due to too low resolution, however, between some signals in one dimensional (1D) ^1H LF-NMR, energy relaxation spectra of T_2 and T_1 of lignocellulose samples, had poor peak resolution [34]. Using sparse representation methods based on PDCO (primal-dual interior method for convex objectives), we recently reported of developing a computing approach for maximization of the efficiency of inverse Laplace transformation of ^1H LF-NMR energy TD relaxation data to spectral graphs [19, 27]. In the present work with some modifications and adaptation, we could

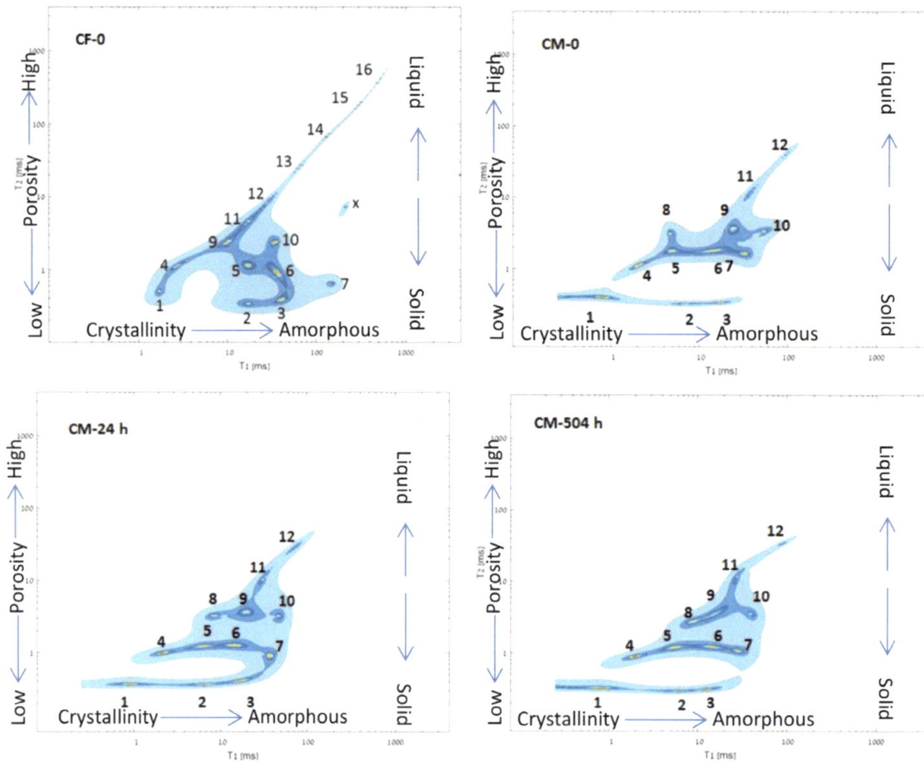

Figure 2.
Two dimensional (2D) chemical and morphology domain spectrum maps of T_1-T_2 energy relaxation signal times for graphic mapping of cattle forage (CF) and cattle manure (CM) for 21 days of AD reaction (CM-0 h; CM-24 h; and CM-504 h). T_1 correlates with the crystallinity/amorphous level of the tested complex components and T_2 correlates with the level of porosity of the complex components. Time-domain peaks 1–10 are solid and time domain peaks and above peak 11 are in liquid phase. Color ladder: Light blue – Low ^{1}H intensity; strong blue – Medium ^{1}H intensity; and yellow – High ^{1}H intensity.

specifically generate by signal reconstruction 2D graphs of T_1-T_2, with significantly improved peak resolution of lignocellulose samples described in detail in a previous study [19]. This allowed us to generate a detailed qualitative analysis of 20% hydrated complex cattle forage (CF) and CM biomass's (**Figure 2**) in morphological and chemical aggregate energy relaxation time domains (TD) and to follow their TD peaks changes during AD conversion of CM to biogas. In agreement with previous reports [13, 14] and our recent work [19], peak identification in the 2D graphs described in subsequent paragraphs, showed that the alkyl chain's ^{1}H and absorbed water's ^{1}H T_1 spin–lattice energy relaxation time changes are influenced by the different chemical structures and crystallinity levels of the cellulosic structure, and T_2 spin–spin changes are in part influenced by the porosity of the solid polymeric layer of lignocellulose. In effect, these two ^{1}H energy relaxation time values (T_1 and T_2) are providing information related to both chemical composition and morphological structure.

The 2D chemical-morphological graphics presented in **Figure 2**, described in detail below covers the following material samples:

A. Sample CF: We initiated the 2D T_1-T_2 graph study with cattle forage (CF) samples that are characterized with minimal AD, and used as a comparative control for the CM samples at different times of AD (**Figure 2** CM 0 h, CM 24 h to CM 504 h). The CF sample contains digestible and at the same basic non-digestible chemical complexed composition as the initial CM of cellulose, hemicellulose, lignin, pectin and fatty compounds, and other components. However, with time during AD, the CM's 2D chemical and morphological spectrum shows different morphological aggregated structures of the biomass material as will be described.

B. Sample CM time zero: The second comparative sample used is fresh CM (**Figure 2**, CM-0 h). This biomass sample is demonstrating the internal cattle AD process on CF that took place in the cattle's digestive tract.

C. Sample CM time 24 h: The third sample (**Figure 2**, CM-24 h) is after 24 h of AD reaction. This sample represents the CM biomass at the end of the peak of intensive biogas production as shown in **Figure 1B**.

D. Sample CM time 504 h: The fourth and last sample (**Figure 2**, CM-504 h) is CM after 21 days of AD process, demonstrating the chemical and morphological structure at the end of the biogas production process.

3.2.2 Peaks assignment

In **Figure 2** of the samples investigated, required an intensive characterization of the CM's lignocellulose component: In effect, we assigned cellulose, hemicellulose, pectin lignin, lipid and other components and their association/localization within this T_1-T_2 map as described below. The individual ^1H energy relaxation time peaks for CM and cattle forage in **Figure 2** have been assigned to different water-pool interactions as a function of material chemical composition and morphological structural domains of the waters sites of absorption sites: For T_1-T_2 peak assignments, we used three approaches as described in the experimental section. The first was based on characterizing individual component standards (lignocelluloses components found in CM, including cellulose; hemicelluloses; pectin and lignin) for their peaks in a 2D T_1-T_2 graphic map. The second was based on spiking the CM or CF sample with the same aforementioned individual known standards. The third approach was based on standard chemical hydrolysis methodologies of cellulose, hemicelluloses, lignin, and pectin.

Peak assignment is also based on previous studies [19, 37–39] of similar CM biomass samples used in the present study. Cellulose is a linear biopolymer consisting of repetitive glucose units, linked by β-1,4 glycocidic bonds. Parallel cellulose chains are then assembled by hydrogen bonds and van der Waals forces into microfibrils, which are often embedded in a matrix of other polymers, such as hemicellulose and lignin. Hemicellulose is a relatively amorphous heterogeneous polysaccharide composed of a series of randomly branched polysaccharides [4, 8, 40], of which the short branch chains usually contain different monomers such as xylose, mannose, galactose, rhamnose, and arabinose. In addition, hemicellulose interacts with cellulose, pectin, and lignin giving a network structure resistant to hydrolysis [4, 28]. Lignin is a large, complex aromatic heteropolymer present in cell walls made of different phenyl propane alcohols. The content and composition of lignin varies between plant species, and functions to increase the strength of plant cell wells. Pectin known as a gelling agent is a structural heteropolysaccharidein the primary cell walls and is particularly abundant in the non-woody parts of terrestrial plants, and its content is relatively high in forage [36].

In our study CF and CM described below, the lignocellulosic biomass is divided into three morphological domains D1 (peak groups 1, 2, 3), D2 (peak groups 4, 5, 6, 7), and D3 (peak groups 9, 10, peak 8) is missing in CF sample and will be discussed later on.

Each domain has similar T_2 relaxation times and is divided to sub-domains, of individual peaks, by the different T_1 values. Other non-lignocellulosic components, such as free sugars and fatty acid compounds, are assigned domains D4 (peak group 11) and D5 (peak groups 12–16), respectively.

Based on the T_1-T_2 mapping strategy for lignocellulose complex materials described above, to characterize internal morphology and chemical structures, it is possible to explain the differences between each T_1-T_2 peaks, and to suggest a rational assignment for the first ten lignocellulosic associated peaks found in the different CF and CM samples studied. A brief peak assignment is listed below:

Peak 1 in **Figure 2** has the fastest, in effect shortest, T_2 and T_1 energy relaxation times. This is rationalized by the smallest matrix pore size (lowest T_2) and thus the strongest water ^1H interaction with the internal surface of the nano-aggregate most crystalized cellulosic domain (lowest T_1). Considering the 2D graphic mapping of cellulose standard [26] and the cellulose specific chemical hydrolysis test, we assigned peak 1 to the cellulose's mostly crystalline interior domain of the lignocellulose fiber complex. This is in good agreement with the state of art for lignocellulose models [4, 28]. For peak 10 with the longest T_2 and T_1 is well associated with the localization to the amorphous cellulose standard as well as spiking tests using this standard [19], which suggests an assignment to the most amorphous cellulose outer surface domain of the lignocellulose fibers that correlates with the elementary cellulose microfibril model [28]. Interestingly, since the forage sample is rich in young vegetable grasses dominated by primary cell wall, and high in pectin [4], it is highly hydrated as it contains relatively higher amounts of amorphous cellulose domains as seen by the peak heights in forage (**Figure 2**, CF) compared to the CM (**Figure 2**, CM).

Peaks 2 and 3 having a longer T_1 than peak 1 can be rationalized as a reduction of water proton interaction strength with part of the interior crystalline cellulose domain's surface consisting of cellulose partially associated with hemicelluloses and pectin chains.

Peak 4 is assigned as a subsurface second layer [28]. It consists of para-crystalline morphology.

Peak 5 fits well to the standard and spiking localization tests of oxidized polyunsaturated fatty acid domain [26]. This standard is rich in aldehyde end groups and will be discussed later.

Peak 6 is assigned to medium-hydrate para-crystalline nano-aggregated cellulose complex [28]. A relatively higher degree of hydration typical for this peak is explained because of the relatively open hemicellulose "loop chains" that may absorb more water [28, 34] in comparison to peak 4.

Peak 7 is assigned to a hydrated partially amorphous cellulose subsurface layer.

Peaks 8, 9: Based on the polyunsaturated fatty oxidized standard localization analysis and data from relevant literature reports [3], peak 8 is assigned to a well-developed outer surface microbial epoxy hydrophobic nano-aggregated domain. These peaks as well as peak 5 are generated not by water ^1H but mainly from aliphatic chains characterized by T_1 equal to T_2. Considering lignocellulose components standard spiking, peak 9 domain is assigned to a semi-hydrated surface amorphous nano-aggregated fraction of depolymerized cellulosic oligosaccharide products, associated with hemicellulose and trapped lignin.

Peak 10: Due to its relatively long T_1 and T_2 energy relaxation times and localization of most amorphous cellulose standards, we assigned peak 10 to a high amorphous packing, and in agreement with the literature [4, 32] to an amorphous cellulosic moiety with an outer hydrated surface on a highly amorphous domain of cellulose of the CM components with the potential for hydrolysis and release of free sugars (glucose), shown in peak 11, as identified by standards and spiking tests [19].

Peak 11: Based on standard and spiking tests, peak 11 is assigned to monosaccharide free sugar subdomains.

Peak 12 is attributed to disaccharide free sugars.

Peaks groups 13–15 appear only in the spectrum map of CF. They are assigned by standards test, to fatty acids with different carbon chain lengths and the alkyl chain's degree of unsaturation. These fatty acids were easily absorbed in the cattle intestine system and/or metabolized by the internal anaerobic bacteria digestion activity.

Based on the above peak assignments: CF sample in **Figure 2** shows the T_1-T_2 relaxation time graphic map of the cattle forage biomass, and is compared to the CM biomass, as a relatively non-anaerobic digested organic control biomass material for visualization of chemical and morphological spectrum changes due to cattle rumen bacterial digestive activity and absorption of cattle feeding material components. It can be seen in **Figure 2** that CF sample is much richer in the free sugars (peak 11) in comparison to CM-0 h. Furthermore, triacylglycerol and all graphic spectrum peaks of conventional vegetable oils and fatty acids (peaks 14–16 shown in **Figure 2**) is obtained only in the CF sample and is totally missing in the graphic spectrum of CM samples. The assignment of all these T_1-T_2 peaks was confirmed by graphic mapping of sugars and fatty acids standards previously reported by us [19] for CM biomass. At the lower part of the diagonal, in **Figure 2** at first view CF, graphic map may show a different pattern of peak distribution that can be rationalized by the metabolism process of CM sample. However, following a deeper view, it can be observed that CF has many basic similarities to CM-0 h. A relatively small pore-size material domain (in effect a low T_2) within the suggested interior of lignocellulose crystalline layer (peaks 1, 2, and 3) can be seen in the **Figure 2** graphic map of CM and CF as well. Other domains represent a second intermediate porosity layer representing the subsurface (peaks 4, 5, 6, and 7) of the elementary cellulose microfibril model [28], followed by a more increased pore-size layer of the fibril surface assigned to peaks 9 and 10.

Interestingly, in CF, T_1-T_2 energy relaxation time graphic map (**Figure 2**, CF) peaks 5, 6, and 7 are partly shifted to increased T_1 (due to higher absorption of water and longer ^1H energy relaxation times) in comparison to CM energy relaxation graphic map (**Figure 2**, CM-0 h). This signal shift to relatively longer T_1 energy relaxation times suggests changes of the molecular environment of the cattle forage nano-aggregated complex may be explained by the significantly higher content of pectin in forage sample [37, 38] known to have a stronger water interactions with gel-like properties (as discussed previously). All these significant spectral T_1-T_2 peak shifts/changes between CF vs. CM samples, indicating that the present 2D relaxation mapping system is able to distinguish between different molecular morphologies and supramolecular interactions. This is in agreement with standards localization and spiking analysis described above for CM, with variations rationally suggested to be assigned to high pectin concentration environments associated with cellulose, since the forage sample is rich in young vegetable grasses dominated by primary cell wall, high in pectin [36] that is highly hydrated and contains relatively high amorphous cellulose domains. Furthermore, peak 8 is missing in forage biomass sample (**Figure 2**, CF) in comparison to CM relaxation map (**Figure 2**, CM-0 h, CM-24 h, and CM-504 h), which is generate by ^1H on alkyl chains present only in CM due to developed anaerobic bacterial biofilm during digestion. Thus some T_1-T_2 peaks identify domains related to materials processed and formed by digestion in the cattle's rumen by anaerobic microbial activity. In effect for CM, the forage biomass underwent an AD process in the cattle rumen.

In the AD process of CM, a significant degradation is observed for cellulose fibers and a significant microbial colonization layer (biofilm) presence, which is assigned to peak 8. In the section on quantitative determination of ^1H proton density per peak, we can show biofilm growth beyond CM-0 h and CF samples, in CM-24 h and 504 h samples. This pattern of both surface cellulose degradation and increase of an extracellular polysaccharide (EPS) thick biofilm layer is expected [19]. The standard localization test of oxidized polymers rich in aldehydes support this material assignment to peak 8 T_1-T_2 at 4.32–2.71 ms.

3.2.3 Monitoring of T1/T2 ratio of CM biomass T_1-T_2 TD

In this study, ^1H T_1 and T_2 energy relaxation times of absorbed water protons are used to characterize different morphological domains in CM and how these domains change during AD. As described above, the different morphological domains within CM are compared, and the changes they undergo, by a 2D relaxation graphic map of T_1-T_2 peaks. Another important aspect we would like to characterize is the interactive strength of the water-proton absorption within the different CM morphologies. In effect with respect to the different degrees of crystallinity, the strength of the water's proton absorption on a given morphology may be correlated to the degree of crystallinity of the substrate. As described in the introduction and further below, one measure of interactive strength is the T_1/T_2 ratio of a given T_1-T_2 peak.

The ^1H relaxation times of absorbed water T_1 are generally larger than T_2 [21], and this is also seen in the present study of CM. In the T_2 vs. T_1 2D graphs (**Figure 2**) of undigested CM (CM-0 h), all peaks are with T_1 equal or larger than T_2, and are thus below the dotted diagonal ($T_1/T_2 = 1$) in the 2D map. As described above, $T_1 = T_2$ relaxation times in bulk water, but in absorbed water the ^1H relaxations are $T_1 > T_2$ and thus $T_1/T_2 > 1$. This can be seen in **Table 2** giving the T_1/T_2 ratio for different morphological sites in the 2D CM T_1-T_2 relaxation graphs at different AD times.

The values of T_1/T_2 in **Table 2** can be correlated with the strength of the interactions between ^1H H_2O and the surface, which can be rationalized in our study by the degree of crystallization and the number of –OH on the cellulose available for interaction with the absorbed water molecule. Thus, the higher the T_1/T_2 value the stronger the interaction for a given morphological domain due to the greater reduction of T_2 than T_1 when going from bulk water to adsorbed water on external and porous surfaces.

One set of interesting results is the comparison between peaks 1, 2 and 3, which vary from highly crystalline to semi-hydrated crystalline respectively, in the T_1-T_2 graphs (**Figure 2**) in terms of T_1/T_2 values. Each of these peaks is associated with a crystalline cellulose domain, wherein the each of these crystalline domains is enclosed in different environments. In this line, the T_2 relaxation values are relatively constants and the T_1 values increase significantly, such that the T_1/T_2 ratio

Time	Peak										
	1	2	3	4	5	6	7	8	9	10	11
0 h	2.21	20.81	54.60	2.00	3.89	9.15	23.16	1.60	7.54	19.92	3.61
24 h	2.20	17.35	37.30	2.25	4.59	12.66	37.06	2.97	4.71	13.49	2.81
504 h	2.51	23.29	38.75	2.38	4.73	9.98	25.04	3.44	4.00	11.10	2.54

Table 2.
T_1/T_2 ratio [ms/ms] of different morphological sites in 2D T_1-T_2 relaxation time of CM.

goes from in the range of 2–20–55 for peaks 1, 2, and 3, respectively. This increase is attributed to increasing T_1 values, brought about by lower interaction strength of the absorbed water ^1H with the local environment which we hypothesis is due to changes in the environment of the crystalline cellulose nano/micro morphologies, which lowers interactive strength and increases T_1. This same trend is seen at all stages of the AD for T_1-T_2 domains 1, 2, and 3. In T_1-T_2, peaks 5, 6, and 7 which are less crystalline and significantly amorphous, wherein the T_2 stays approximately constant the same trend is seen with increasing T_1 values with the T_1/T_2 ratio going from 4 to 9 to 23, respectively.

If we look at peaks 3, 6, and 9 (crystalline interior, semi-hydrated sub surface, and amorphous surfaces, respectively), which go from crystalline to less crystalline to highly amorphous morphologies, wherein the T_1 is relatively constant the T_1/T_2 ratio decreases from 55, 9, and 7.5 at time zero which at time 21 days goes from 38.7, 10 to 4, the trend in T_1/T_2 is similar and indicates that T_1/T_2 is changing because of the increasing strength of interaction with the given morphology, which appears to be the strongest with the more crystalline (peak 3) and decreasing with the increase in the more amorphous morphologies and different chemical compositions such as hemicellulose and lignin. This reduction in the strength of the interaction may correspond to the increases in T_2 in the amorphous regions.

If we look at T_1-T_2 domains 7, 10, and 11 (amorphous sub-surface, highly amorphous surface, and free sugars, respectively) with a relative similar T_1, the T_2 values increase and thus T_1/T_2 ratios decrease, which may be due to decrease in the interactive strength of adsorbed ^1H water interactions with the surrounding environment due to decreases in cellulosic chains and thus decreasing hydrogen bonding interactions.

If T_1 value and T_2 both increase by the same factor—the ratio T_1/T_2 does not change. For example, peaks 7 (amorphous sub-surface) and 3 (crystalline interior, described above in Section 3.2.2) have high and similar pattern of T_1/T_2 ratio from 24 to 504 h (in **Table 2**), indicating similar strengths of water adsorption though differences in morphology, wherein other components in the surrounding environments may affect the results by competitive interactions with the ^1H interactions.

3.3 SEM of CM biomass fibers changes during AD

To further support the conclusions on chemical-morphological arrangements in CF and CM, we did SEM analysis to visually observe changes of biomass fibers during 21 days of mesophilic anaerobic reaction (AD). In **Figure 3**, degradation of cellulose fibers (Black arrows) is readily seen at magnification ×1000 and ×5000. Developed bacterial biofilms are seen at magnification ×20,000 (**Figure 3**, black dashed arrows): In the forage, there are only isolated bacterial colonies most likely on most of the amorphous surface. In the two low magnification SEM photos (×1000 and ×5000), clear intact cellulosic fibers can be easily seen (**Figure 3**). In the relatively higher SEM magnification of ×20,000, small and non-developed and non-continuous spots/layer of bacterial biofilm could be observed. This biofilm can be explained by the silage process that the forage passed before being used as CF.

In the sample of CM-0 h, partially digested lignocellulose fibers are clearly seen and a significant bacterial biofilm layer covering the surface is seen in all three SEM magnification images.

The SEM images of CM-24 h show a similar pattern of fibers digestion and a thicker growing continuous bacterial biofilm layer that covers the entire surface. Following 504 h (21 days) of AD, the lignocellulose fibers seems to be most degraded in the two low SEM magnification (×1000 and ×5000) and the biofilm layer is also observable as in the sample of CM-24 h.

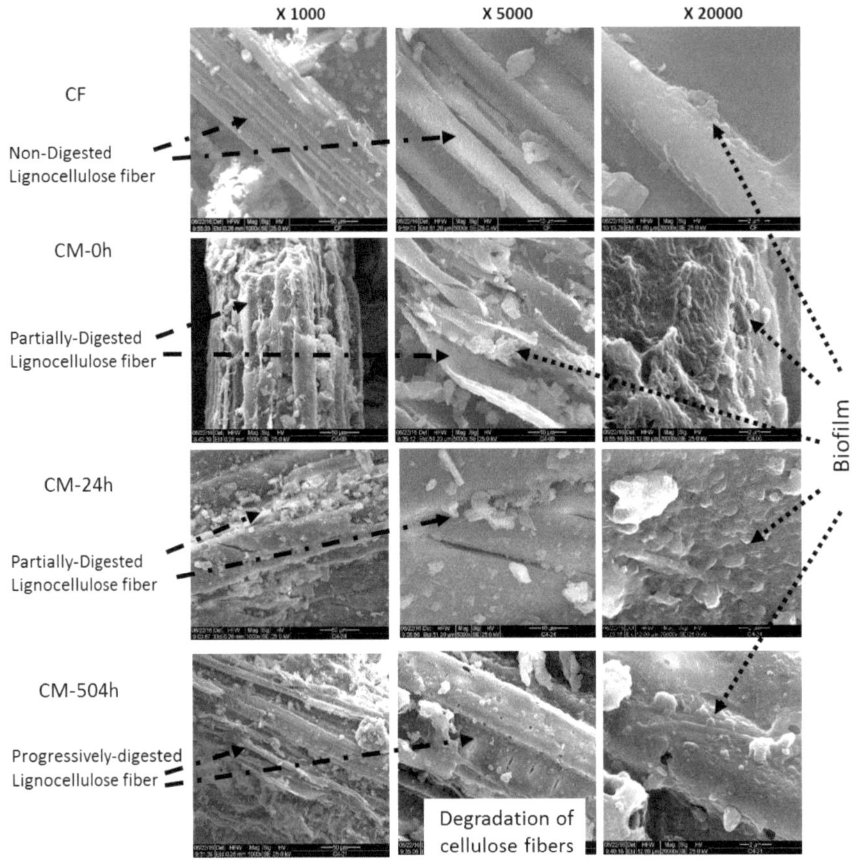

Figure 3.
Scanning electrons microscope (SEM) images of cattle forage (CF) and cattle manure (CM) at different times of anaerobic digestion (AD): 0, 24, and 504 h. different stages of lignocellulose digestion are shown by arrows in left side and anaerobic bacterial biofilm is shown by arrows in the right side of the figure.

3.4 Quantitative monitoring of accessible amorphous cellulosic components during 21 days of mesophilic anaerobic reactions

To carry out a quantitative analysis of the results based on the 2D T_1-T_2 relaxation study: We developed a computing protocol (described in details in supplemental section S2), which enabled the quantification of proton (^1H) concentrations within all peak domains measured during the anaerobic digestion process (**Table 3**).

The results show that proton populations of peaks 1 and 4 assigned as most crystalline cellulose nano-aggregated complexes are stable during most of the digestion process, while peak 1 increased during the last period near to 21 days (**Table 3**).

The relative volumes of the most amorphous peaks 7 and 10 are considerably reduced, which we suggest to be rationalized is due to the fact that these two amorphous morphologies are most readily subjected to bacterial degradation. The proton population of peaks 5 and 8, assigned to oxidized aggregated structures of bacterial EPS biofilm (shown in S3), show a general increase from the initial stage to the end of AD process at 21 day (**Table 3**).

Proton population of peaks 11, assigned to free sugars that are continuously released from the amorphous fractions of the nano-aggregated cellulosic complexes and then degraded to volatile fatty acids to yield the final biogas product, have a relatively stable steady ^1H state concentration during the anaerobic process (**Table 3**).

Time	Peak										
	1	2	3	4	5	6	7	8	9	10	11
0 h	20.58	8.17	6.69	21.54	6.16	13.13	14.12	2.07	2.71	4.38	0.62
24 h	21.58	11.57	9.46	16.66	8.65	11.57	10.82	3.23	6.47	2.16	1.06
504 h	25.94	10.15	8.31	21.76	9.74	9.13	7.05	5.73	6.63	1.32	1.26

Table 3.
A peak's relative volume (%) compared to all the peaks within a 3D T_1-T_2 relaxation map of cattle manure during anaerobic digestion.

3.5 Description of the amorphous cellulose component's peak 10 and its correlation with equivalent changes of CM with biogas production rate during 21 days of AD

As a follow up of our previous report [19] demonstrating the usefulness of peak 10 for monitoring of amorphous cellulose equivalent (ACE), we now correlated it with biomethane production rate during 21 days of mesophilic anaerobic reaction. Peak 10 has the longest T_1 and T_2 relaxation times of all others lignocellulosic domains. Indicating that this domain is composed of the most loosely packed and less ordered aggregate structure, such that water molecules have lower interaction forces with the morphological surfaces allowing higher water mobility during energy exchange with cellulose fibers surface. Furthermore, during AD, this peak 10 domain being amorphous allows better access of the water solubilized hydrolytic enzymes. Hence, the domain represented by peak 10 are more readily degradable suggesting it is readily degraded to free sugars that are further hydrolyzed to short chain volatile fatty acids such as acetate, propionate, butyrate that are used by methanogenic bacteria to generate methane.

Taking advantage of the ability to quantify T_1-T_2 relaxation signals, with respect to ^1H concentration, a good correlation between peak 10's volumes with the biogas flow was achieved (**Figure 4**). Within the first 24 h, when biogas flow significantly increases to a maximum level, peak 10's volume decreases to a lower level compared to the original values. Along the entire period of 21 days of the AD process, this inverse correlation of peak volume to gas flow is clearly demonstrated. This finding enables us to suggest that peak 10 as an indicator for digestibility of cellulosic components in complex chemical and morphological mixtures.

As was already demonstrated in our previous paper [19], the signal volume of amorphous cellulose at peak 10 shows a highly linear fit in our 2D T_1-T_2 system.

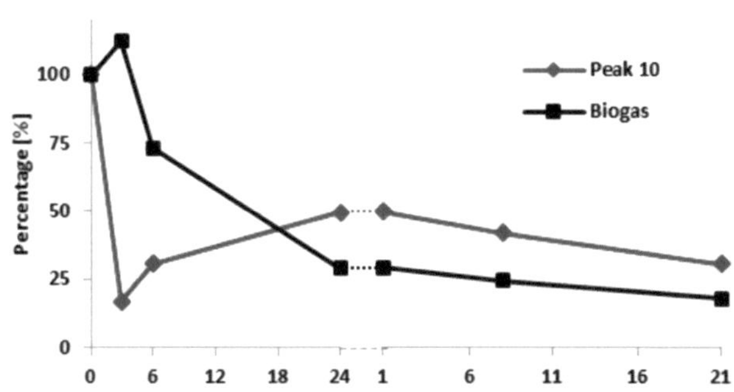

Figure 4.
Relative biogas flow rate and signal volume of peak 10 during 21 days of AD. Left side of the curve is the first 24 h of AD and the right side is from 24 to 504 h. there is an inverse match between the volume of peak 10 and the biogas flow suggesting peak 10 is the amorphous cellulose degraded straight forward to biogas.

Time	Peak 10 signal [Intensity (AU)]	ACE in a sample (mg)	Total ACE (% Dry)	Total ACE in digester (g)
CF	2655	483	16.1	—
CM 0 h	748	136	4.53	117
CM 3 h	133	24.2	0.81	20.8
CM 6 h	250	45.4	1.51	39.0
CM 24 h	410	74.5	2.48	62.6
CM 192 h	332	60.4	2.01	47.6
CM 504 h	264	48.0	1.60	32.7

Table 4.
Amorphous cellulose equivalent (ACE) of cattle manure during CM anaerobic digestion and of forage. (AU is in arbitrary units).

Thus indicating every 5.5 signal units of peak 10 in the 2D relaxation map are equivalent to 1 mg of amorphous cellulose. Therefore, we can calculate the amorphous cellulose equivalent (ACE) mg of cellulosic biomass as shown in **Table 4**.

The concentration of ACE in forage (CF) is much higher than in CM because it mostly raw none-degraded biomass. It is clearly obtained in **Table 4** in term of signal intensity (represented by volume) of peak 10 is dramatically high (X3.5) in the control CF sample than in CM 0 h sample. During the first hours of AD process of CM, peak 10 intensity significantly declines after which until the end of the process at 21 days the signal intensity mildly fluctuated. Similar pattern of fluctuation is found in absolute values (mg) and also in percentage of ACE. It should be noticed that although ACE of CM is reduced during AD, it keeps a low value due to constant degradation of the inner layer of lignocellulose that creates new amorphous regions. The total ACE of CM before digestion was 117 g and immediately decreases to its minimum value of 20.8 g due to a peak in bacterial activity and biogas production then slightly increases and fluctuates for the rest of the digestion period due to consumption-production fluctuation rates.

4. Conclusion

We demonstrated the monitoring of the chemical and morphological changes of CM biomass during anaerobic mesophilic digestion in a biogas production process, by T_1-T_2 energy relaxation time graphs generated by signal reconstruction from LF ^{1}H NMR. Graphical peak changes in 2D LF ^{1}H NMR monitored the rate of crystallinity and porosity changes of the CM lignocellulosic biomass, and provided a better understanding of its conversion to biogas mechanism. This demonstrated the versatile usage of 2D ^{1}H NMR T_1-T_2 technology in studying complex chemical and morphological systems, with the practical potential for improving biomass conversion efficiency into biomethane.

In addition to accurate qualitative mapping of the different energy relaxation time domains (TD) and subdomains of the different nano-aggregates in CM biomass complexes, a detailed peak assignment to chemical and morphological structures was formulated. The proton NMR relaxation TD sensor provided also quantitative data of the changes of the key amorphous cellulose morphology that is accessible to bacterial digestive enzymes to release free sugars that are fermented by methanogenic bacteria to produce the final product of bio-methane and carbon dioxide. Furthermore, the facile NMR TD sensor also provides chemical and morphological information on the growth and changes of bacterial biofilm during the anaerobic digestion (AD) of the CM biomass. This information may be used to study optimal preconditioning of samples to maximize bio-methane yield and to identify optimal termination times of the AD process.

Acknowledgements

We would like to thank Dr. O. Levi and Dr. D. Benson from the Open University for the mathematical and statistical backup and support of this study. We would also like to thank Mr. Natan Ayallon, Ms. Maysa Resende, Dr. J. Abramovich, Ms. S. Kravcik, and all the members of the Phyto-Lipid Biotechnology Lab (PLBL) for the intensive technical assistance. This study was supported by a grant from the Israeli Ministry of Science Technology and Space.

Competing financial interests

The authors declare no competing financial interests.

Author details

Wiesman Zeev* and Linder Charles
Phyto-Lipid Biotechnology Laboratory (PLBL), Department of Biotechnology Engineering, Energy Engineering Unit, Faculty of Engineering Sciences, Ben Gurion University of the Negev, Beer Sheva, Israel

*Address all correspondence to: wiesman@bgu.ac.il

References

[1] Chynoweth DP, Owens JM, Legrand R. Renewable methane from anaerobic digestion of biomass. Renewable Energy. 2001;**22**:1-8

[2] Holm-Nielsen JB, Al Seadi T, Oleskowicz-Popiel P. The future of anaerobic digestion and biogas utilization. Bioresource Technology. 2009;**100**:5478-5484

[3] Yu Z, Morrison M, Schanbacher FL. Production and utilization of methane biogas as renewable fuel, chapter 20. In: Vertès AA, Blaschek HP, Yukawa H, Qureshi N, editors. Biomass to Biofuels: Strategies for Global Industries. Hoboken, NJ, USA: John Wiley & Sons; 2010. pp. 403-433

[4] Appels L, Lauwers J, Degrève J, Helsen L, Lievens B, Willems K, et al. Anaerobic digestion in global bio-energy production: Potential and research challenges. Renewable and Sustainable Energy Reviews. 2011;**15**:4295-4301

[5] Dumitrache A, Wolfaardt GM, Allen DG, Liss SN, Lynd LR. Tracking the cellulolytic activity of *Clostridium thermocellum* biofilms. Biotechnology for Biofuels. 2013;**6**:1-14

[6] Fernández N, Díaz EE, Amils R, Sanz JL. Analysis of microbial community during biofilm development in an anaerobic wastewater treatment reactor. Microbial Ecology. 2008;**56**:121-132

[7] Langer S, Schropp D, Bengelsdorf FR, Othman M, Kazda M. Dynamics of biofilm formation during anaerobic digestion of organic waste. Anaerobe. 2014;**29**:44-51

[8] Chen S, Liao W, Liu C. Value-Added Chemicals from Animal Manure. Richland, WA, USA: Pacific Northwest National Laboratory; 2003. pp. 1-135

[9] Taherzadeh MJ, Jeihanipour A. Recalcitrance of lignocellulosic biomass to anaerobic digestion, chapter 2. In: Mudhoo A, editor. Biogas Production: Pretreatment Methods in Anaerobic Digestion. New York, USA: Wiley; 2012. pp. 27-54

[10] Ding S, Himmel ME. The maize primary cell wall microfibril: A new model derived from direct visualization. Journal of Agricultural and Food Chemistry. 2006;**54**:597-606

[11] Quiroz-Castañeda RE, Folch-Mallol JL. Hydrolysis of biomass mediated by cellulases for the production of sugars, chapter 6. In: Quiroz-Castañeda RE, Folch-Mallol JL, editors. Sustainable Degradation of Lignocellulosic Biomass- Techniques, Applications and Commercialization. Rijeka: InTechOpen; 2013. pp. 119-155

[12] Bernardinelli OD, Novotny EE, de Azevêdo ER, Colnago LA. Analyses of biomass products by nuclear magnetic resonance spectroscopy, chapter 6. In: Vaz Jr. S, editor. Analytical Techniques and Methods for Biomass. Cham: Springer International Publishing; 2016. pp. 143-172

[13] Chandel AK, Antunes FA, Anjos V, Bell MJ, Rodrigues LN, Polikarpov I, et al. Multi-scale structural and chemical analysis of sugarcane bagasse in the process of sequential acid–base pretreatment and ethanol production by *Scheffersomyces shehatae* and *Saccharomyces cerevisiae*. Biotechnology for Biofuels. 2014;**7**:1-17

[14] Gómez X, Diaz MC, Cooper M, Blanco D, Morán A, Snape CE. Study of biological stabilization processes of cattle and poultry manure by thermogravimetric analysis and 13 CNMR. Chemosphere. 2007;**68**:1889-1897

[15] Park S, Baker JO, Himmel ME, Parilla P, Johnson DK. Cellulose crystallinity index: Measurement techniques and their impact on interpreting cellulase performance. Biotechnology for Biofuels. 2010;**3**:1-10

[16] Silverstein RM, Webster FX, Kiemle DJ, Bryce DL. Carbon-13 NMR spectrometry, chapter 4. In: Spectrometric Identification of Organic Compound. New York, USA: John Wiley & Sons; 2014. pp. 204-244

[17] Hills B, Benamira S, Marigheto N, Wright K. T 1-T 2 correlation analysis of complex foods. Applied Magnetic Resonance. 2004;**26**:543-560

[18] Song Y, Venkataramanan L, Hürlimann M, Flaum M, Frulla P, Straley C. T1–T2 correlation spectra obtained using a fast two-dimensional Laplace inversion. Journal of Magnetic Resonance. 2002;**154**:261-268

[19] Wiesman Z, Linder C, Resende MT, Ayalon N, Levi O, Bernardinelli OD, et al. 2D and 3D Spectrum graphics of the chemical-morphological domains of complex biomass by low field proton NMR signal analysis. Energy & Fuels. 2018;**32**:5090-5102

[20] McDonald P, Korb J, Mitchell J, Monteilhet L. Surface relaxation and chemical exchange in hydrating cement pastes: A two-dimensional NMR relaxation study. Physical Review E. 2005;**72**:011409

[21] D'Agostino C, Mitchell J, Mantle MD, Gladden LF. Interpretation of NMR relaxation as a tool for characterising the adsorption strength of liquids inside porous materials. Chemistry - A European Journal. 2014;**20**:13009-13015

[22] Berman P, Meiri N, Colnago LA, Moraes TB, Linder C, Levi O, et al. Study of liquid-phase molecular packing interactions and morphology of fatty acid methyl esters (biodiesel). Biotechnology for Biofuels. 2015;**8**:1-16

[23] Cox J, McDonald PJ, Gardiner BA. A study of water exchange in wood by means of 2D NMR relaxation correlation and exchange. Holzforschung. 2010;**64**:259-266

[24] Foston M, Ragauskas AJ. Changes in the structure of the cellulose fiber wall during dilute acid pretreatment in Populus studied by 1H and 2H NMR. Energy & Fuels. 2010;**24**:5677-5685

[25] Meng X, Foston M, Leisen J, DeMartini J, Wyman CE, Ragauskas AJ. Determination of porosity of lignocellulosic biomass before and after pretreatment by using Simons' stain and NMR techniques. Bioresource Technology. 2013;**144**:467-476

[26] Berman P, Levi O, Parmet Y, Saunders M, Wiesman Z. Laplace inversion of low-resolution NMR relaxometry data using sparse representation methods. Concepts in Magnetic Resonance Part A. 2013b;**42**:72-88

[27] Campisi-Pinto S, Levi L, Benson D, Cohen M, Resende MT, Saunders M, et al. Analysis of the regularization parameters of primal-dual interior method for convex objectives applied to 1H low field nuclear magnetic resonance data processing. Applied Magnetic Resonance. 2018;**49**:1129-1150

[28] Berman P, Leshem A, Etziony O, Levi O, Parmet Y, Saunders M, et al. Novel 1 H low field nuclear magnetic resonance applications for the field of biodiesel. Biotechnology for Biofuels. 2013a;**6**:1-20

[29] Berman P, Meiri N, Linder C, Wiesman Z. 1H low field nuclear magnetic resonance relaxometry for probing biodiesel autoxidation. Fuel. 2016;**177**:315-325

[30] Reid JD, Lynch D. Cellulose Analysis. Industrial and Engineering Chemistry, Analytical Edition. 1937;**9**:570-573

[31] Rico C, Rico JL, Tejero I, Muñoz N, Gómez B. Anaerobic digestion of the liquid fraction of dairy manure in pilot plant for biogas production: Residual methane yield of digestate. Waste Management. 2011;**31**:2167-2173

[32] Gerardi MH. The Microbiology of Anaerobic Digesters. Hoboken, NJ, USA: John Wiley & Sons; 2003. pp. 1-192

[33] Nasir IM, Mohd Ghazi TI, Omar R. Anaerobic digestion technology in livestock manure treatment for biogas production: A review. Engineering in Life Sciences. 2012;**12**:258-269

[34] Ono H, Shimaya Y, Sato K, Hongo T. 1H spin-spin relaxation time of water and rheological properties of cellulose nanofiber dispersion, transparent cellulose hydrogel (TCG). Polymer Journal. 2004;**36**:684-694

[35] Felby C, Thygesen LG, Kristensen JB, Jørgensen H, Elder T. Cellulose–water interactions during enzymatic hydrolysis as studied by time domain NMR. Cellulose. 2008;**15**:703-710

[36] Froix MF, Nelson R. The interaction of water with cellulose from nuclear magnetic resonance relaxation times. Macromolecules. 1975;**8**:726-730

[37] Poletto M, Ornaghi HL, Zattera AJ. Native cellulose: Structure, characterization and thermal properties. Materials. 2014;**7**:6105-6119

[38] Morán JI, Alvarez VA, Cyras VP, Vázquez A. Extraction of cellulose and preparation of nanocellulose from sisal fibers. Cellulose. 2008;**15**:149-159

[39] Vermaas JV, Petridis L, Qi X, Schulz R, Lindner B, Smith JC. Mechanism of lignin inhibition of enzymatic biomass deconstruction. Biotechnology for Biofuels. 2015;**8**:1-16

[40] MacKay AL, Wallace JC, Sasaki K, Taylor IE. Investigation of the physical structure of the primary plant cell wall by proton magnetic resonance. Biochemistry. 1988;**27**:1467-1473

Section 2

Solution and Solid State NMR

Chapter 4

Molecular Interactions in Ionic Liquids: The NMR Contribution towards Tailored Solvents

Mónica M. Lopes, Raquel V. Barrulas, Tiago G. Paiva, Ana S.D. Ferreira, Marcileia Zanatta and Marta C. Corvo

Abstract

Ionic liquids have been on the spotlight of chemical research field in the last decades. Their physical properties (low vapor pressure, thermal stability, and conductivity) and the possibility of fine tuning make them a versatile class of compounds for a wide range of applications, such as catalysis, energy, and material sciences. Ionic liquids can establish multiple intermolecular interactions with solutes such as electrostatic, van der Waals, or hydrogen bonds. The prospect of designing ionic liquid structures toward specific applications has attracted the attention to these alternative solvents. However, their rational design demands a molecular detailed view, and Nuclear Magnetic Resonance is a unique and privileged technique for this purpose, as it provides atomic resolution and at the same time enables the study of dynamic information. In this chapter, we provide an overview about the application of Nuclear Magnetic Resonance spectroscopy techniques as a methodology for the rational design of ionic liquids as solvents for small organic compounds, CO_2 capture, and polymers such as cellulose focusing mainly in the last 10 years.

Keywords: ionic liquids, nuclear magnetic resonance spectroscopy, molecular interactions, chemical shift deviations, nuclear Overhauser effect, solute/solvent

1. Introduction

Ionic liquids (ILs) can be defined as ionic species that melt under 100°C. These materials are usually made from organic cations such as imidazolium, ammonium, pyrrolidinium, or phosphonium, and organic or inorganic anions such as chloride, bromide, tetrafluoroborate, or bis(trifluoromethylsulfonyl)imide (**Figure 1**). Their physical properties, namely low vapor pressure, thermal stability, and conductivity grant them the status of alternatives to organic solvents in a wide range of applications, from catalysis and energy applications to material sciences. These properties are a direct consequence of the identity and interactions between the cations and the anions; however, the vast number of possible combinations hampers the tailoring of ILs as solvents for specific applications. In order to establish the desired structure-property relationships, a molecular understanding on structure and

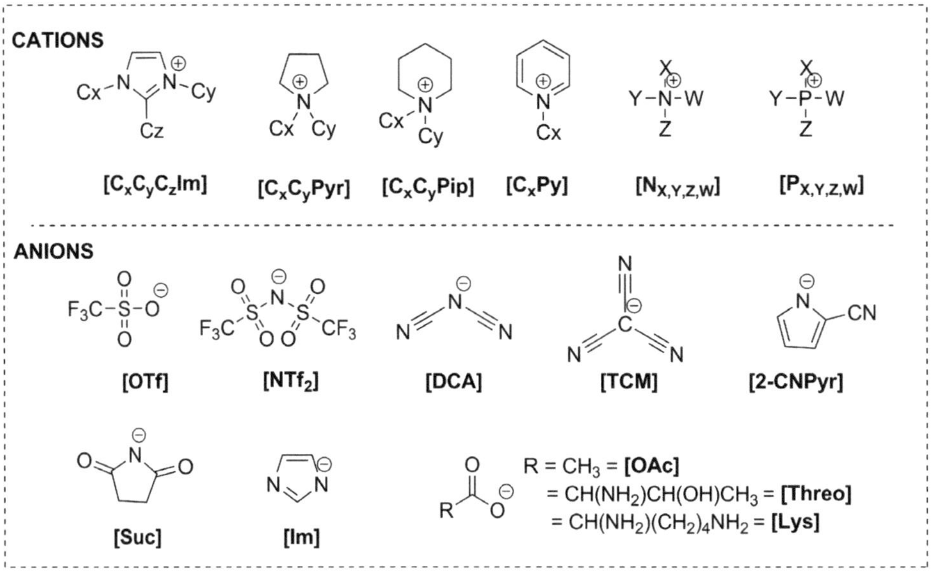

Figure 1.
Common cation and anion structure of ILs. The abbreviations used in this chapter are described in Section 7.

interactions is required. Nuclear Magnetic Resonance (NMR) techniques enable this study by allowing to probe both cations and anions through several nuclei (^{1}H, ^{13}C, ^{19}F, ^{35}Cl, ^{11}B, ^{15}N, and ^{31}P). In addition, the use of chemical shift deviations, relaxation, NOE, and diffusion experiments allows advanced studies about cation-anion-solute interactions and consequently facilitates the design and IL applications.

As solvents, ILs can establish several attractive interactions that range from weak, nonspecific, and isotropic forces such as van der Waals, solvophobic, and dispersion forces, to strong interactions as Coulombic forces and anisotropic forces such as hydrogen bonds (H-bonds), halogen bonds, dipole-dipole, magnetic dipole, and electron pair donor/acceptor interactions. The diversity and strength of intermolecular forces in ILs are responsible for the local arrangements in bulk and interface [1].

NMR enables the study of structure and dynamics in ILs providing information that allows the rationalization of the solvation behavior [2, 3]. Thus, by obtaining the profile of IL molecular interactions, it is possible to select and optimize the cation and anion identity and their relation to solvated species. In particular, the use of multinuclear NMR experiments and homo and heteronuclear correlation protocols enables the *in-situ* evaluation of cation-cation, cation-anion, anion-anion, and also IL-solute relationships.

In neat imidazolium ILs (ImILs), H-bonds are formed between the cation and the anion. This attractive force in which hydrogen nuclei are bound to electronegative atoms such as nitrogen, oxygen, or fluorine is mainly felt by the aromatic protons of the imidazolium cation (H_2, H_4, and H_5) and establishes a regular network. When a H-bond is formed, the hydrogen's electron density is pulled by an electronegative atom, leading to a deshielding of this nucleus and consequent positive chemical shift deviation ($\Delta\delta$). If a H-bond is weakened or broken, the opposite occurs, leading to a negative $\Delta\delta$. The addition of a solute to the IL can establish additional H-bonds and also introduce perturbations to the regular network, either reinforcing the interaction between cation and anion (positive $\Delta\delta$) or weakening (negative $\Delta\delta$) (**Figure 2I**) [4, 5].

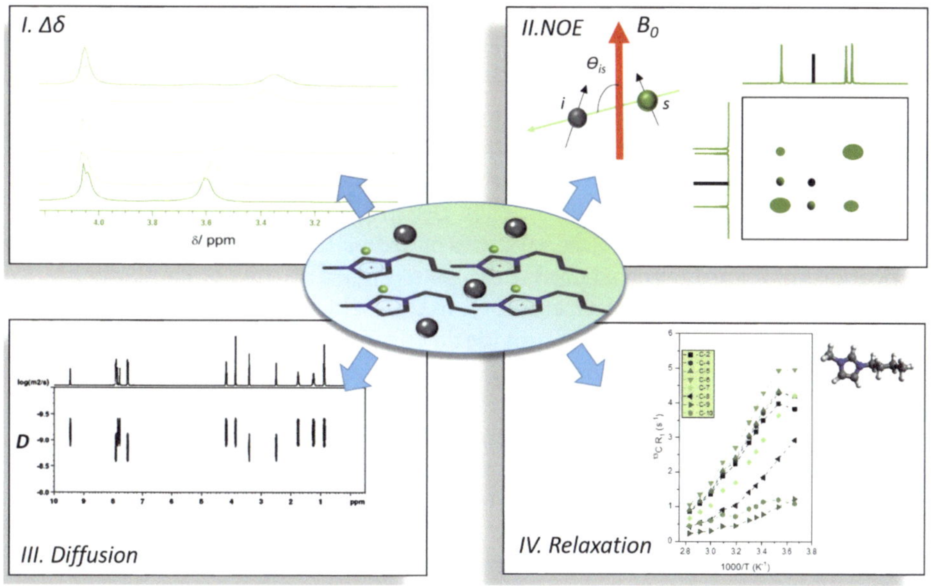

Figure 2.
NMR experimental techniques used to probe molecular interactions in ILs.

The nuclear Overhauser effect (NOE) is commonly used to study local structure in liquids and solutions. Considering the two nuclear spins, *I* and *S*, the magnetization transfer through coupling of their magnetic dipole moments produces a cross-relaxation and a consequent NOE (**Figure 2II**). Increasing the internuclear distance (r), the cross-relaxation rate decays with $1/r^6$. Thus, a NOE is only felt if r is less than 4–5 Å. Although this theory still holds for intramolecular distances, a new concept was introduced by Gabl et al. to analyze intermolecular NOE in ILs. According to these authors, the distance rule is strongly dependent on the frequencies of the interacting nuclei and can instead vary with $1/r$. Site-specific NOE should be interpreted cautiously as it can reflect the mean orientation of the ions over longer distances, rather than the local structure of distinct ion aggregates. Nevertheless, homonuclear (NOESY) and heteronuclear NOE (HOESY) studies have been widely used to analyze intermolecular interactions of solutes with ILs [6–8].

In ILs, the self-diffusion coefficients in solution can be measured through diffusion ordered spectroscopy (DOSY) providing a means to understand charge transport properties. In particular, using pulsed field gradient NMR (PFG-NMR), the spatial positions of the nuclei are encoded by magnetic field gradients, in which the signal intensity attenuation is correlated with the diffusion time (Δ) and the gradient-related parameters (g, δ) [9]. Multinuclear diffusion measurements can give access to anion and cation self-diffusion coefficients (**Figure 2III**); however, since ion-pair interactions take place on a timescale faster than the diffusion, the coefficients are a weighted average of charged and neutral species [10].

A deeper insight into atomic level interactions comes from NMR relaxation techniques (**Figure 2IV**). NMR spin-lattice (T_1) and spin-spin (T_2) relaxation measurements allow the study of reorientation dynamics in ILs and the analysis of inter- and intramolecular interactions [11–13].

Overall, probing intermolecular interactions in ILs through simple 1D NMR spectrum analysis ($\Delta\delta$), or more complex relaxation phenomenon, NOE experiments, and diffusion provides an understanding of the interplay between ILs and solutes. The following sections highlight several examples where NMR techniques were used to tailor ILs.

2. Interactions between ionic liquids and small organic molecules

The optimization of liquid-liquid extraction processes for small organic molecules with organic solvents usually requires high investment and large energy consumption due to unfavorable capacity and selectivity [14]. Since this process is common in many industrial areas, more sustainable procedures are required in order to reduce organic solvent volatiles. Usually, an efficient separation of azeotropic systems containing the mixtures of aromatic/saturated hydrocarbons or alcohols/aliphatic hydrocarbons is difficult to obtain because of similar boiling points and several possible azeotropes. In this field, ILs have been cited as an alternative solvent, since they can provide an enhanced solvation capability while having negligible vapor pressure [14, 15]. The efficiency of ILs for absorption of aromatic compounds is determined by the size and structure of both cations and anions, as well as the molecular structure of the aromatic compounds. This section will focus on IL-aromatic molecule interactions consisting of two main groups: (i) unsubstituted and (ii) substituted aromatic molecules.

2.1 Unsubstituted aromatic molecules

Kumar and Banerjee reported that smaller IL cations enhanced the extraction capability for aromatic compounds [16]. Later, Potdar et al. studied binary mixtures of $[C_2C_1Im]^+$ with $[C_1SO_4]^-$, $[C_2SO_4]^-$, and $[C_1SO_3]^-$ as counterions to extract benzene from hexane, measuring the equilibrium compositions by ^{1}H NMR spectroscopy [14]. $[C_2C_1Im][C_2SO_4]$ achieved higher selectivity and capacity than IL binary mixtures. The combination of two ILs could result in lower free volume than a single IL, and benzene molecules would be unable to penetrate the core of the solvent.

Other studies in 2014 by Dias et al. reported interactions between benzene and $[C_2C_1Im][NTf_2]$, through several NMR experiments, namely, 1D ^{1}H and ^{13}C spectroscopy and 2D ^{1}H,^{1}H-NOESY spectroscopy [17]. In the mixture of benzene with $[C_2C_1Im][NTf_2]$, H_2 presented a lower chemical shift (more shielded) than neat IL, which indicated that the cation was positioned above and below the aromatic ring plane, and the anion was located in the same plane as benzene protons [17, 18]. Early works on the absorption of aromatic sulfur compounds by Su et al. studied the interaction between thiophene and $[C_4C_1Im][PF_6]$, $[C_4C_1Im][BF_4]$, and $[C_2C_1Im][BF_4]$, through 1D ^{1}H, ^{19}F, ^{11}B, and ^{31}P spectroscopy [19]. The ring current effect in neat thiophene was stronger than in neat $[C_4C_1Im][PF_6]$, because in the former, molecules were tightly packed, whereas in the IL, $[C_4C_1Im]^+$ cations were separated by $[PF_6]^-$ anions. The bulky $[PF_6]^-$ prevented $[C_4C_1Im]^+$ cations from coming close and reduced the aromatic current effect. In $[C_4C_1Im][PF_6]$/thiophene mixture, an upfield shift in $[C_4C_1Im]^+$ protons was explained as a result of the CH-π interaction between $[C_4C_1Im]^+$ and thiophene aromatic ring. The trend of $\Delta\delta$ indicated that $[C_4C_1Im]^+$ was in the shielding cone of thiophene ring current and the thiophene was located near the deshielding zone of $[C_4C_1Im]^+$ ring current. In addition, thiophene is not a good H-bond acceptor because the unpaired electrons of sulfur participate in the aromatic ring current and may carry a partial positive charge. The $[PF_6]^-$ anion was a potential hydrogen acceptor but introduced a downfield shift. The dilution of $[C_4C_1Im][PF_6]$ by a neutral molecule like thiophene was not expected to break apart the strong Coulombic interaction between $[C_4C_1Im]^+$ and $[PF_6]^-$ ions significantly by a simple insertion between them. The electric field effect risen from the physical interaction between fluoro-based anions and cations or thiophene was not a dominant factor in determining the IL absorption capacity for thiophene. The upfield shift effect was strongest for the ring protons in the imidazolium cation followed by the methyl and methylene protons next to the nitrogen. The further away the protons were from the ring as in 1-butyl or 1-ethyl groups, the weaker the upfield shift effect became. Therefore, thiophene molecules must locate close to the ring of the IL's imidazolium cation. The CH-π interaction between the acidic hydrogen of cation and thiophene was not as important as the π (imidazolium ring)-π (thiophene ring) interaction. Apparently, the aromatic ring current effect (π-π interaction) is the dominant factor to determine the Δδ trend in these IL systems. The chain length of the imidazolium alkyl group and the size and character of anion affected the coordination number and geometric arrangement of thiophene molecules relative to IL's cations and anions. The stacking structure of the cation and anion was tightest for the neat [C_2C_1Im][BF_4], followed by [C_4C_1Im][BF_4] and [C_4C_1Im][PF_6]. Therefore, the negative charge of [BF_4]$^-$ anion in the [C_2C_1IM][BF_4] was spread more around fluorine atoms and less around boron than those in the [C_4C_1Im][BF_4] system. It also suggested that the [BF_4]$^-$ anion in the [C_2C_1Im][BF_4] system is more basic and interacts more intimately with the imidazolium cation than in [C_4C_1Im][BF_4]. The NMR results suggested that the partial positive charge in thiophene sulfur should direct the anions outside the thiophene plane. Overall, the aromatic ring current interaction between the imidazolium cations and the thiophene was responsible for the absorption of aromatic sulfur compounds in ILs [19].

Revelli et al. studied the interaction of thiophene with [C_1C_1Im][(C_1O)PHO_2] and [C_4C_1Im][SCN], using ^{1}H and ^{31}P NMR spectroscopy [15]. The dilution of [C_4C_1Im][SCN] by thiophene did not lead to a visible ^{1}H Δδ; therefore, thiophene did not break apart the strong Coulombic interactions between cation and anion. Indeed, phase equilibria of [C_4C_1Im][SCN]/thiophene system indicated that thiophene had low solubility in this IL. The acidic hydrogen atoms of the imidazolium cation formed weak H-bonds with the sulfur of thiocyanate anions. ^{1}H NMR chemical shifts of the ring protons in neat thiophene were more upfield than those in neat [C_4C_1Im][SCN], indicating, as previously, that the aromatic ring current effect was stronger in neat thiophene than in neat [C_4C_1Im][SCN]. The bulky [SCN]$^-$ anion prevented the [C_4C_1Im]$^+$ cation from coming close, thus greatly reducing the aromatic current effect. Similar studies were made to unveil the interactions between thiophene and [C_1C_1Im][(C_1O)PHO_2]. The Δδ of thiophene and [C_1C_1Im][(C_1O)PHO_2] followed the same trend as in [C_4C_1Im][SCN]. The chemical shifts of thiophene protons were more downfield when the thiophene is in a mixture of IL, although the IL proton chemical shifts were upfield. The ^{31}P chemical shifts were measured at several molar ratios of thiophene in [C_1C_1Im][(C_1O)PHO_2], but no Δδ was observed; hence, the interactions between thiophene and anion are weak. The most prominent features in the vicinity of thiophene are an excess of cations in the polar regions above and below the ring and an excess of anions around the equator of thiophene. In the end, [C_4C_1Im][SCN] is a promising IL to act as a solvent for desulfurization [15].

Kaintz et al. contributed to the measurements of solute diffusion in IL solutions through infinite dilution, since it eliminates solute-solute interactions. The authors studied the interactions between [C_nC_1Pyr][NTf_2] (n = 3, 4, 6, 8, and 10) and [$P_{6,6,6,14}$][NTf_2] and naphthalene, anthracene, pyrene, and biphenyl, through pulsed field gradient ^{1}H NMR. Within the selected set of aromatic solutes and pyrrolidinium ILs, no dependence was found on solute dipole moment. Neither was there any evidence that the polar/nonpolar domain formation in the higher homologs of the [C_nC_1Pyr][NTf_2] series or in [$P_{6,6,6,14}$][NTf_2] influences diffusion of solutes as small as benzene. Through Stokes-Einstein predictions, it was possible to observe that self-diffusion in ILs is like self-diffusion in conventional solvents. However, the relation to solute/solvent size ratio was greater in ILs than in conventional solvents [10].

2.2 Substituted aromatic molecules

2.2.1 Polyphenols (Catechins)

Barrulas et al. gave an important contribution for the NMR study of interactions between ILs and small molecules [20]. The articles so far were not very complete in terms of the NMR experiments available, lacking always some proof of concept. Extracts from plants like matcha green tea have considerable significance as bioactive compounds with several pharmacological applications, namely polyphenols (being catechins the most prevalent). Nonetheless, the amount of these compounds in the extracts is typically very low. Consequently, green extraction techniques with higher efficiency for phenolic compounds are of paramount importance, in which ILs can be used as designer solvents. In order to rationalize the interactions, several model compounds were selected – naphthalene to study π-π stacking, veratrole, L-3,4-dihydroxyphenylalanine (L-DOPA) to understand the behavior of the phenol moiety toward H-bonds and hydrophobic interactions, and 1,2-dihydroxyanthraquinone (alizarin) in order to study all the interactions simultaneously. Several ILs were tested: $[C_4C_1Im]Cl$, $[BzC_1Im]Cl$, $[C_2C_1Im]Cl$, $[C_4C_1C_1Im]Cl$, $[C_2C_1Im][OTf]$, $[C_2C_1Im][TCM]$, $[C_2C_1Im][DCA]$, and $[C_4C_1Im][OTf]$. The solutions resulting from the mixture of the ILs and the model compounds were tested through ^{1}H-DOSY and ^{1}H,^{1}H-NOESY NMR experiments (**Figure 3**). Both the IL cation and anion had an influence on the solvent behavior.

The best ILs for extracting matcha polyphenols were imidazolium derivatives with shorter alkyl sidechains and weakly basic anions such as $[TCM]^-$, $[DCA]^-$, and $[OTf]^-$. Finally, it was possible to locate π-π stacking and hydrophobic interactions in case of the cation and H-bond acceptor interactions in case of the anion [20].

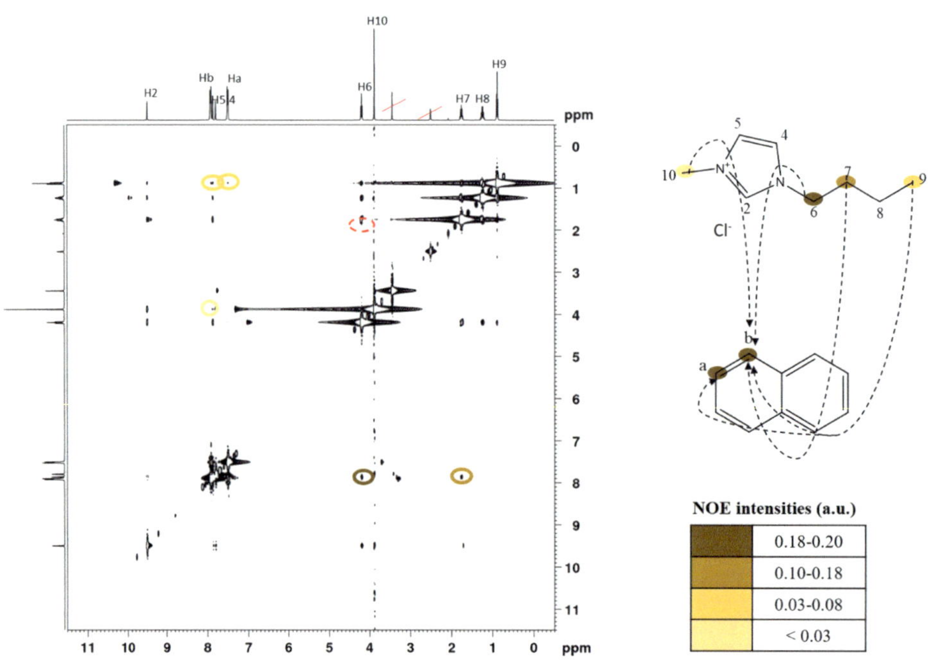

Figure 3.
2D ^{1}H,^{1}H-NOESY spectrum with 200 ms mixing time of $[C_4C_1Im]Cl$/naphthalene mixture [0.16:0.16 (M/M) in DMSO-d_6] and main intermolecular interactions (adapted from Ref. [20]).

2.2.2 5-Hydroxymethylfurfural

Nowadays, searching for new sources of green energy is crucial. Biomass is receiving attention, since the reserves are abundant, the price is low and is renewable. The preparation of 5-hydroxymethylfurfural (5-HMF) from biomass is a very interesting topic, since this compound is applied in the production of high value-added chemicals and biofuels [5, 21]. Thus, new high-efficiency reaction medium for the preparation of 5-HMF is needed, and ILs are suitable alternatives. Mixtures of 5-HMF and ILs with anion $[NO_3]^-$ and different cations $[C_4C_1Im]^+$, $[C_4Py]^+$, $[C_4C_1Pyr]^+$, $[C_nC_1Pip]^+$, and $[N_{1,1,1,H}]^+$ were studied and presented strong intermolecular interactions. In order to unveil the chemical nature of such interactions, ^{1}H NMR experiments were done for different ILs/5-HMF systems, and the Δδ of each proton was estimated [5]. In $[C_4Py][NO_3]$, the H-bond between the anion and the cation was mainly formed by O atom in $[NO_3]^-$ and H_2, H_5 atoms that were close to the N atom of the pyridine ring [5, 22]. Therefore, the upfield movement of H_2 and H_5 protons indicated that the input of 5-HMF into ILs weakens the H-bonding between anion and cation. With the addition of 5-HMF into ILs, the π-π interactions between pyridine rings became weaker and the shielding effect was strengthened, causing the protons of the alkyl chain in cation upfield shift. Based on ^{1}H,^{1}H-NOESY experiments, it was seen that the addition of 5-HMF induced the change of alkyl chain conformations, from gauche to trans. With the exception of $[C_4C_1Pyr]^+$ and $[C_nC_1Pip]^+$, $[C_4C_1Im]^+$, $[C_4Py]^+$, and $[N_{1,1,1,H}]^+$ could form H-bonds with hydroxyl group in 5-HMF, and this capacity increased as follows: $[C_4Py]^+ \approx [C_4C_1Im]^+ <$ $[N_{1,1,1,H}]^+$. 5-HMF could strongly bind IL anions, while the H-bond interaction of 5-HMF with the cation was much weaker. Also, the interactions between the carbonyl group of 5-HMF and the cation of ILs were also very weak. Therefore, the cation impacted the interaction of 5-HMF with ILs through its interaction with the anion [5].

2.2.3 Fluorinated derivatives of benzene

Dias et al. studied the interactions between mixtures of $[C_2C_1Im][NTf_2]$ with benzene (discussed in Subsection 2.1), fluorobenzene, 1,2-difluorobenzene, 1,4-difluorobenzene, 1,3,5-trifluorobenzene, 1,2,4,5-tetrafluorobenzene, pentafluorobenzene, and hexafluorobenzene through 1D ^{1}H NMR and ^{13}C NMR and 2D ^{1}H,^{1}H-NOESY NMR spectroscopy [17]. Relative to ^{1}H NMR, it is important to highlight that proton H_2 was the most interactive site (most acidic proton) of the imidazolium cation, which means that any perturbation in the polar network of the IL was felt at this position. In case of the benzene mixtures, the aromatic protons exhibited larger shielding shifts than their aliphatic counterparts; thus, in case of $[C_2C_1Im][NTf_2]$/1,2-difluorobenzene mixtures, the H_2 protons exhibited small shielding shifts as the aromatic molecule concentration increased. At higher concentrations, these turned into small deshielding shifts. The aliphatic protons showed intermediate shielding shifts, and the H_4/H_5 protons displayed the largest shielding shifts. In case of $[C_2C_1Im][NTf_2]$/hexafluorobenzene mixtures, most shifts were deshielding through the order of $H_2 > H_6\text{–}H_8 > H_4/H_5$. When the aromatic molecules are interspersed in the polar network of the IL, there is a competition between the interactions with the oxygen atoms of the $[NTf_2]^-$ anion (OBT) and the π electrons or fluorine atoms of the aromatic solute. The polar network was stronger at the H_2–OBT contacts than at the H_4/H_5–OBT contacts. This led to a higher proportion of H_4 and H_5 protons interacting with the solute molecules, relative to H_2 interactions. With ^{13}C NMR, it was possible to observe that the carbon atoms attached to hydrogen atoms experienced small shielding shifts in the benzene mixtures that changed to larger deshielding shifts in mixtures with more fluorinated aromatic molecules. The carbon atoms linked to fluorine atoms experienced deshielding shifts in the hexafluorobenzene mixtures that became progressively larger shielding shifts in mixtures with less fluorinated aromatic molecules. 2D NOESY spectra of the equimolar mixtures of [C_2C_1Im][NTf_2] with 1,4-difluorobenzene and pentafluorobenzene identified a cross-peak between the H_8 protons of the IL and the protons of the aromatic molecule. Summing up, in benzene, the cations are mainly located above and below the aromatic plane, experiencing the diamagnetic influence of the aromatic electrons. As the aromatic ring becomes much more fluorinated, the cations migrate to the equatorial plane of the aromatic molecule, to a region of milder paramagnetic effect.

3. Interactions between ionic liquids and CO_2

NMR spectroscopy allows the direct measurement of CO_2 solubility as well as the evaluation of intermolecular interactions and sorption mechanism [23, 24]. Since ILs exhibit the ability to tailor many of their physical and chemical properties, such as gas solubility, they have been extensively studied in CO_2 sorption. Different sorption mechanisms can be observed according to the site and the strength of CO_2/IL interaction. Therefore, according to the CO_2 sorption mechanism, the ILs can be divided into three main groups: physisorption, chemisorption, and water sorption (**Figure 4**).

In general, CO_2/anion interactions are preferential in comparison to CO_2/cations; therefore, the classification of ILs in these three groups is mainly based on the anion functional group and respective IL properties. Non-basic nucleophilic anions capture CO_2 through a noncovalent pathway, and these anions belong to the first group. Basic anions that can form covalent bonds with CO_2 molecule belong to the second and third groups. Besides the functional groups of the anions and the mechanism of sorption, the main difference between ILs of the second and third groups is the products that are formed, since the third group reacts with the water existing in the IL.

Furthermore, the pressure and temperature may also affect CO_2 solubilization. For lower pressure applications, the chemisorption is favored; meanwhile, increasing the pressure, physisorption becomes preferential. A smart selection of pressure and IL structure is essential to develop a higher CO_2-philic material. It is important to

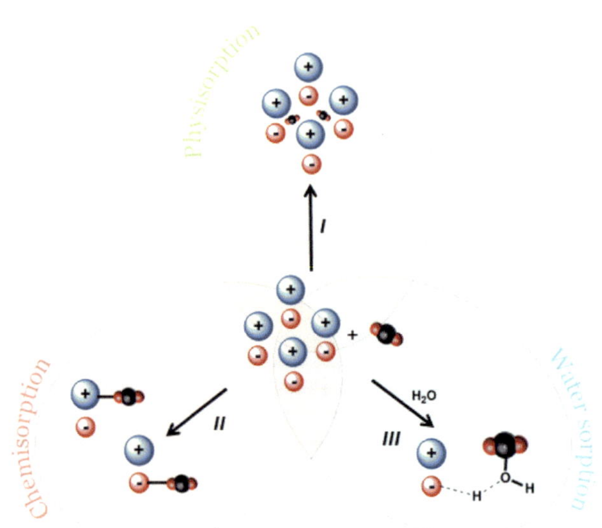

Figure 4.
Summary of CO_2-IL interactions observed by NMR experiments.

DOI: http://dx.doi.org/10.5772/intechopen.89182

mention that all CO_2 capture mechanisms can occur simultaneously in a single IL. In this context, the NMR spectroscopy allows direct measurement of CO_2 solubility, intermolecular interactions, and sorption mechanism and performs *in situ* high pressure experiments. Therefore, the NMR is a powerful technique for designing ILs with high steric hindrance and low viscosity to capture CO_2 efficiently and selectively [23].

3.1 Noncovalent interactions—physisorption

To design an efficient sorbent for CO_2 capture, CO_2 transport properties need to be understood. In physisorption, the surface area of IL attracts CO_2 molecules by van der Waals forces. In this case, CO_2 sorption occurs mainly in ILs containing nonbasic-nucleophilic anions, like fluorinated anions. Based on 3D structures of ImILs that create ion cage-like structures with void spaces capable of confining CO_2, several NMR techniques were being used to study the capability of ImILs to absorb CO_2 and understand IL/CO_2 interactions.

Corvo and co-workers studied $[C_4C_1Im][PF_6]$, $[C_4C_1Im][BF_4]$, and other fluorinated ILs with slight changes in the aliphatic substituent as chain branching or elongation, such as $[(i\text{Pr})C_2C_1Im][NTf_2]$ and $[C_4C_1Im][NTf_2]$, respectively. The cation/anion, cation/cation, cation/CO_2, and anion/CO_2 interactions were analyzed through NOE, DOSY, and relaxation experiments, at 10 and 80 bar, combined with molecular dynamics (MD) simulations (**Figure 5**) [24, 25].

The comparison of ^1H-^1H NOE cross-peak intensity in the presence and absence of CO_2 showed the direct cation/cation interactions. Cross-peaks between the methyl group and the methylene of the aliphatic chain are particularly informative because they reflect only intercationic interactions. From the NOESY data for

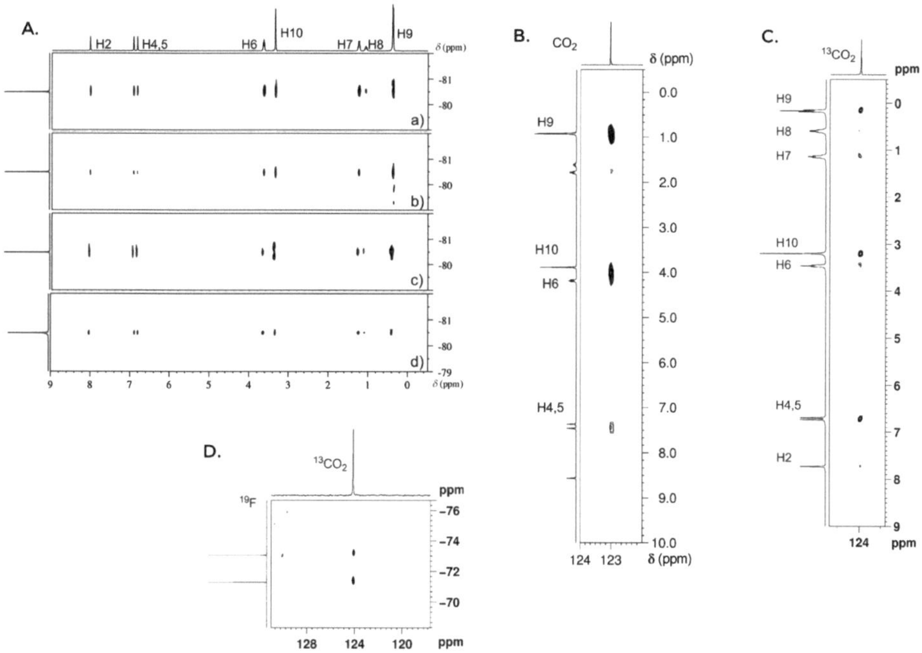

Figure 5.
(A) ^1H,^{19}F-HOESY spectra with 400 ms mixing time of $[(iPr)C_2C_1Im][NTf_2]$: (a) neat IL at 298 K; (b) mixed with 10 bar CO_2 at 298 K; (c) neat IL at 313 K; (d) mixed with 80 bar CO_2 at 313 K; (B) ^{13}C,^1H-HOESY spectrum of $[(iPr)C_4C_1Im][NTf_2]$, with 10 bar of $^{13}CO_2$ at 298 K; (C) ^{13}C,^1H-HOESY spectra (500 ms mixing time) of $[C_4C_1Im][PF_6]$, mixed with $^{13}CO_2$ at 298 K and 11 bar; (D) ^{13}C,^{19}F-HOESY (400 ms mixing time) of $[C_4C_1Im][PF_6]$, mixed with $^{13}CO_2$ at 298 K and 11 bar (adapted from Refs. [24, 25]).

neat IL and IL/CO_2 solutions, it was possible to conclude that at 80 bar, the pattern of cation/cation interactions changes considerably with the correlations between protons of the methyl group H_{10} (NCH_3) and the sidechain protons $H_7/_8$ almost absent from the spectra.

^1H,^{19}F-HOESY experiments showed essentially the cation-anion contacts. Due to the geometry of the anion, the $[NTf_2]^-$ in ImIL was near to the terminal methyl groups (RCH_3 and NCH_3) in linear and branched chain. After CO_2 dissolution, the similarity on the cross-peak intensity from the ^1H,^{19}F-HOESY experiment indicated that the cation/anion contacts were mostly the same as in the neat ILs (**Figure 5a**). The main differences between the neat ILs and the IL/CO_2 mixtures indicated the locations that were mostly affected by CO_2 solvation. At 10 bar of CO_2, the aromatic protons H_4/H_5 presented stronger correlations with the anion and also the methyl group NCH_3. At 80 bar of CO_2, the fluorine contacts with the nonpolar domains, specifically with the methylene protons and the methyl groups, decrease slightly. As **Figure 5b** demonstrates, in the ^{13}C,^1H-HOESY spectra of $[(iPr)C_2C_1Im][NTf_2]/CO_2$ mixture, CO_2 was located around the methyl group RCH_3 and the isopropyl methyl group, proving the direct cation-CO_2 interactions. The HOESY experiments also indicate that increasing CO_2 pressure may lead to the deformation of the ion cage structure of the ILs, accommodating more CO_2 molecules.

^{13}C,^1H-HOESY and ^{13}C,^{19}F-HOESY experiments proved cation-CO_2 and anion-CO_2 interactions. Analyzing the ^{13}C, ^1H-HOESY spectra from the $[C_4C_1Im][PF_6]$, at 10 bar using ^{13}C-labeled CO_2 was possible to notice strong interactions between CO_2 and the aromatic protons H_4/H_5, the methyl group RCH_3 of alkyl chain and the methyl protons of NCH_3 (**Figures 5b, c**). In the same way, the ^{13}C,^{19}F-HOESY experiment confirmed the existence of direct anion-CO_2 interaction (**Figure 5d**).

By using ^{13}C-T_1 NMR relaxation experiments, it was possible to conclude that at low pressure, the solubilization of CO_2 has a minimum effect in the global mobility of the IL. Meanwhile, the opposite was observed for higher pressure (80 bar). Taking into consideration all the experimental data, it was possible to conclude that the physisorption of CO_2 occurs in the cavities near methyl groups NCH_3 and RCH_3 and aromatic protons H_4/H_5 [24, 25].

Taking advantage of known physical sorption of CO_2 in $[C_4C_1Im][NTf_2]$, Allen and Damodaran had chosen this IL to evaluate the diffusion of CO_2 in ILs. Slice-selective inverse gated pulse field gradient technique was employed and demonstrated as an excellent alternative to study gas dynamic solvation under non-equilibrium conditions in a method similar to magnetic resonance imaging (MRI) [26].

3.2 Covalent bond—chemisorption

Many ILs have previously been reported as physical sorbents for CO_2, under different pressures. More recent studies described the capability of ILs to interact through a chemical pathway with CO_2, under low pressures. ILs that contain reactive functional groups, usually N or O moieties on the cation or the anion, can bond covalently with CO_2, which results on the formation of carbamates or carboxylates, respectively. 1D and 2D NMR techniques can easily prove the chemical CO_2-IL interactions.

3.2.1 Carboxylates

Due to the possibility of chemical sorption between CO_2 and the 2-position of the imidazolium ring, ImILs with carboxylate anions have been intensely studied. In 2012, through the ^{13}C NMR analysis, the chemical reaction and the formation of 1,3-dialkylimidazolium-2-carboxylate and acetic acid were reported for the first time, by Blath and co-workers and Danten and co-workers. Both proposed a mechanism of the CO_2 absorption [27–29].

Danten and co-workers started their study with the comparison between the ^{13}C NMR spectra of the pure IL with the CO_2-IL mixture and observed a new signal ca. 155 ppm connected with the CO_2.

The comparison of the ^{15}N NMR spectra of the pure IL with the IL/CO_2 mixture proved the carboxylation into imidazolium ring by the observation of two intense signals at 183.7 and 171.6 ppm assigned to the nitrogen atoms of the ring core. In the mixture, two additional resonance signals flanking each of the previous peaks were observed at 182.1 and 170.0 ppm, respectively (**Figure 6a**). The 2D $^1H,^{15}N$-HMBC spectra of the mixture proved the correlation between the new signals and the secondary peak observed in the ^1H NMR spectrum of the IL/CO_2 mixture (**Figure 6b**).

To understand the favored chemical site among the three possibilities (C_4/C_5 and C_2), the 2D spectrum using the INADEQUATE sequence was performed. The C_2 signal of the $[C_4C_1Im]^+$ cation (138.3 ppm) was not correlated with the two new peaks located at 155.3 and 141.6 ppm (**Figure 6d**). These two peaks were correlated showing the formation of a new species in which CO_2 is covalently bonded to the C_2 of the imidazolium ring (**Figure 6c**) [28, 29].

Another similar study lead by Kortunov, employing non-aqueous $[C_2C_1Im]$ [OAc], supported these conclusions. Once again, the presence of the signal at 155.3 ppm in the ^{13}C NMR spectrum indicated the presence of the carboxylate complex. To reinforce this idea, the $^1H,^{13}C$-HSQC and ^{13}C DEPT-135 NMR experiments were executed, neither the new carbon peaks (141.5 and 155.1 ppm) have protons directly attached nor the new proton signal (15.7 ppm) has any carbon attached [30].

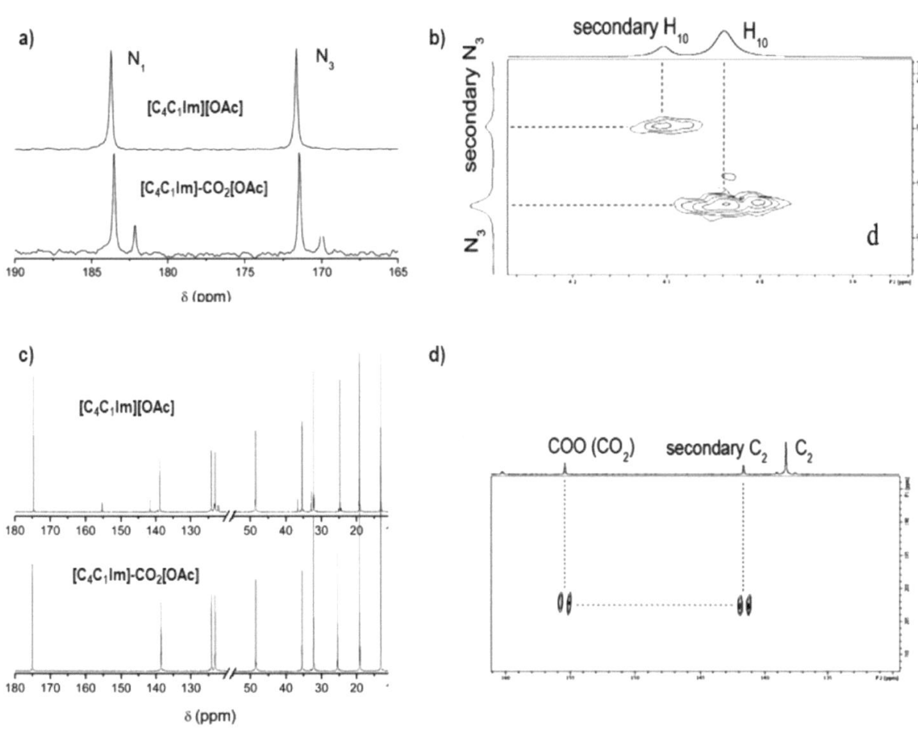

Figure 6.
(a) ^{15}N NMR spectra of the neat $[C_4C_1Im][OAc]$ and $[C_4C_1Im][OAc]/CO_2$ mixture; (b) correlation in $^1H,^{15}N$-HMBC of $[C_4C_1Im][OAc]/CO_2$ mixture; (c) ^{13}C NMR spectra of the neat $[C_4C_1Im][OAc]$ and $[C_4C_1Im]$ $[OAc]/CO_2$ mixture; (d) ^{13}C NMR INADEQUATE of $[C_4C_1Im][OAc]/CO_2$ mixture (adapted from Ref. [28]).

Other essential work using carboxylate anions has been published by Umecky and co-workers. The authors prepared task-specific ImILs with 2,4-pentanedionate and fluorine derivatives to use as CO_2 chemical absorbents. The experiments were performed in neat IL and in dichloromethane solution (50 vol%) by bubbling CO_2. A comparison of ^{1}H NMR spectra among the fluorinated and carboxylate anions showed that the acidity of the hydrogen atom at the 2-position strongly influences the CO_2 sorption capacity. The chemisorption of CO_2 is governed not only by the anion basicity but also by the acidity of H_2 and the possibility of its abstraction. By using 2D analysis (HMQC and HMBC), it was possible to prove the formation of a complex between $[C_2C_1Im]^+$ and CO_2. In those conditions, the influence of dissolved water and consequently the formations of the carbonate (CO_3^{2-}) and bicarbonate (HCO_3^-) were negligible [31].

Ether-functionalized ImILs were also capable of capturing the CO_2 physically and chemically. The physisorption was proved by Seo et al. and the chemisorption by Sharma et al. Both studies were accomplished through ^{13}C NMR and FTIR techniques. Sharma et al. proposed a mechanism for the formation of the carboxylic acid based on ca. 205 ppm signal in the ^{13}C NMR [32, 33].

3.2.2 Amines

The use of ILs containing a basic amine group (task-specific ILs) to capture CO_2 by chemical reactions was reported for the first time by Davis group [34]. After that, amine-based ILs have been the focus of many works, due to their highly efficient capacity on gas solubilization. Generally, the aqueous solution of primary and secondary amines reacts with CO_2 yielding carbamates. On the other side, the reaction between CO_2 and the tertiary amine depends on water amount in the system. The formation of carbamate and bicarbonate has been demonstrated; however, the reaction is still a critical point of discussion and merits further investigations [35, 36].

Combining the idea of low toxicity and high reactivity toward CO_2, a series of choline-based amino acid ILs were reported by Bhattacharyya and Shah. DFT calculations, IR, and NMR spectroscopic techniques were employed to examine the mechanism of the CO_2 interaction with ILs. The ^{1}H, ^{13}C, and ^{1}H,^{1}H-COSY NMR analysis of $[N_{1,1,6,2O4}]$[Lys] demonstrated the signals shifted during the sorption procedure. In ^{13}C NMR spectrum, three new signals were observed: ca. 178 ppm attributed to carboxylate anion with a strong intermolecular hydrogen bonding; at 160.9 and 163.7 ppm attributed to the formation of carbamic acid and carbamate anion, respectively [37].

The CO_2 dissolution and the ionic mobility by diffusion experiments of choline-based ILs $[N_{1,1,6,2O4}]^+$ with $[Threo]^-$ and $[Im]^-$ anions were investigated by the same group. Once again, the formation of carbamic acid and carbamate anion ([Threo]–CO_2) was observed after CO_2 capture, by the presence of characteristic peaks (160.4 and 164.0 ppm) in ^{13}C NMR spectrum. In addition, the $[Threo]^-$ alpha proton represented by a doublet at 3.11 ppm shifted to 3.43 ppm, whereas the beta protons at 3.94 ppm shifted to 4.16 ppm and the methyl proton from 1.20 to 1.29 ppm after CO_2 reaction. By contrast, the ^{1}H NMR spectrum of $[N_{1,1,6,2O4}]$[Im] did not show significant $\Delta\delta$ after CO_2 absorption. However, in the ^{13}C NMR spectrum, a new signal at 160.8 ppm was observed and attributed to the formation of a carbamate specie. Both neat ILs demonstrated broadened signals in ^{1}H NMR spectra, particularly after CO_2 absorption. Using the diffusion decays of the stimulated spin echo, it was possible to observe that the diffusion of anion and cation was not affected by CO_2. The diffusion behavior of both ILs before and after absorbed CO_2 was also studied in a temperature range of 20–90°C. The diffusion of cations and anions changes by the same factor regardless of temperature. From the observed

^1H diffusometry results, the mobility increased for $[Im]^-$ IL after CO_2 sorption and decreased for $[Threo]^-$ anion. The authors explained this behavior based on the decrease in the aromaticity and electron density volume of the imidazolate by the formation of carbamate due to CO_2 reaction [38].

Over the last few years, NMR techniques have proven that the reaction between CO_2 and 1,3-dialkyl imidazolium cation is strongly dependent on the anion [27]. Chen et al. demonstrated that some functionalized ILs can interact physical and chemically with CO_2, by studying the behavior of $[aC_2C_1C_1Im][BF_4]$ with this gas. After CO_2 capture, the ^1H NMR spectrum showed duplicate signals and the ^{13}C NMR spectrum showed two new peaks, proving that part of $[aC_2C_1C_1Im][BF_4]$ interacts with CO_2 chemically, producing carbamic acid IL-NHCOOH (ca. 163 ppm), and the remaining $[aC_2C_1C_1Im][BF_4]$ interacts with CO_2 physically (ca. 160 ppm) [39].

More recently, the influence of CO_2 sorption in ion mobility in aqueous solution (50/50 wt%) of similar choline-based amino acid ILs was evaluated by Filippov et al. Similar to their previous work, they proposed a mechanism of CO_2 sorption in IL containing amino acid anion (such as $[Threo]^-$) process by carbamic acid and carbamate anion formation, proved by the presence of the signal ca. 159 ppm. While the water diffusivity was not affected by the absorption of CO_2, the increase in ion mobility upon CO_2 chemisorption was demonstrated by employing ^1H and ^{13}C NMR diffusion experiments [40].

Brennecke's group prepared a series of alkyl-phosphonium cation ILs with aprotic heterocyclic anions (AHAs) and studied the CO_2 capture in dry and wet conditions. All anions correspond to a tertiary amine incorporated into a heterocyclic ring: indazolide, imidazolide, pyrrolide, pyrazolide, and triazolide-based anions. In dry conditions, they suggested a formation of carbamate species, via Lewis acid-base interaction. Comparing the ^1H NMR chemical shifts of the dry with wet protonated-heterocyclic anions, they concluded that the $[P_{6,6,6,14}]$[2-CNPyr] did not suffer from the anion reprotonation in the presence of water and CO_2, and the opposite trend was previously reported by them for $[P_{6,6,6,14}]$[Pro]. In that case, due to reprotonation and deactivation of the anion, the water reduced the CO_2 molar capacity. Even though they did not discard the bicarbonate formation from CO_2 and water, they believed that the amount of bicarbonate formed was not expected to be prevalent enough at such low CO_2 pressure. For the $[P_{6,6,6,14}]$[2-CNPyr], the addition of 5 wt% of water increased the CO_2 solubility, and the authors suggested a change in the IL-CO_2 complex activity upon water addition. However, no ^{13}C NMR spectra have been reported to confirm the theory [41, 42].

Based on the concept of preorganization, a novel CO_2 sorption mechanism using $[P_{4,4,4,2}]$[Suc] was addressed by Huang et al. Through a combination of ^{13}C NMR, FTIR, and quantum chemical calculations, they suggested an innovative structure organization with superior CO_2 capacity resulting from the interaction between CO_2 and the $[Suc]^-$ anion [43]. Following this idea, the authors also published the tuning of those imide-based anion IL with phosphonium cations by the addition of the electron-donating group. By changing the CO_2 sorption amount, several ^1H NMR and ^{13}C NMR spectra were acquired, and the cooperative interactions between anion and CO_2 were reinforced. According to the authors, at the first stage, the absorption was dominated by N-CO_2 interaction confirmed by the presence of a new signal ca. 159 ppm and by the shift of the signal from 195.6 to 188.0 ppm related to C═O anions. The second stage of absorption mechanism was dominated by the additional O-CO_2 interactions, as suggested by the shift of the same signal to 182.8 ppm. The physical interaction occurred at the last stage and is weaker than the first two. This was confirmed by ^1H NMR spectrum through the shift of the signal from 1.23 to 1.30 ppm during the uptake of CO_2, that indicates the physical hydrogen bond between methylene in the anion and oxygen atom of the CO_2 (C-H-O) [44].

3.3 CO_2 interaction in wet ionic liquids

Since most ILs are hygroscopic, dry ILs are extremely difficult to obtain. The presence of water in ILs can affect many properties, including CO_2 sorption. In this field, a few groups started to evaluate the CO_2 capture in wet ILs. Using analogous phosphonium-based ILs and experimental conditions, our group has demonstrated a different point of view of Huang et al. based on NMR analysis. Combining the ^{13}C-$\{^1H\}$, attached proton test (APT), phase-sensitive HSQC, HMBC, and ^{31}P-$\{^1H\}$ NMR spectra, the degradation of $[Suc]^-$ anion was proposed. The ^{13}C-$\{^1H\}$ NMR spectra showed a second set of signals that did not correspond to the $[P_{4,4,4,2}]^+$ or $[Suc]^-$ at 34.4, 35.4, 174.8, and 176.4 ppm. Those signals were observed at APT analysis and represent two CH_2 groups and two C═O, indicating IL degradation or contamination. Combined with the correlation between protons and carbons observed by HMBC and HSQC experiments, the analysis suggested the formation of 4-amino-4-oxobutanoate (anion from succinamic acid) resulting from the basic hydrolysis of the pyrrolidine-2,5-dione (SucH) during the synthesis of IL. Furthermore, the formation of bicarbonate by the reaction between CO_2 and IL confined water was proposed due to the presence of the signal ca. 159 ppm. The same chemical shift was observed for $[P_{4,4,4,2}]$ $[HCO_3]$ IL, reinforcing the idea of bicarbonate formation [45].

Davis and co-workers demonstrated that the CO_2 capture is hugely impacted by the presence of water in a series of $[N_{1,1,1,1}]$ ILs having amino acid anions. Employing ^1H and ^{13}C NMR analysis, they demonstrated the formation of a carbamate only as transitory specie (ca. 163 ppm) for amino acid ILs. Indeed, when the amount of CO_2 present in the system exceeds about 0.5 mol/mol of IL present, the carbamate complex was detected in trace amounts and the carbonate/bicarbonate capture emerged as predominant. Due to rapid proton exchange, carbonate and bicarbonate appear as a single peak ca. 159 ppm. The study also reported the CO_2 sequestration in wet $[N_{1,1,1,1}]$[2-CNPyr]; in the ^{13}C NMR spectrum, the migration correlated with proton sequestration, and the carbonate/bicarbonate peak was observed; however, the signal associated with carbamate formation was not observed, in opposition of Brennecke's work [41, 42, 46].

Recently, Filippov et al. using solid-state ^{13}C and ^{15}N MAS NMR spectroscopy demonstrated the CO_2 absorption in $[N_{1,1,n,2OH}]$[Threo] and $[N_{1,1,n,2OH}]$[Tau] aqueous solution (50 wt%) ILs. After CO_2 reaction, a precipitated product appeared and was analyzed by ssNMR (^{13}C and ^{15}N), suggesting that the solid sediment is composed of neutral threonine (or taurine) in the zwitterionic forms. Therefore, they assigned the resonance line at 164.4 ppm to the formation of the carbamate anion and the resonance line at 160.6 ppm to the carbamic acid [47].

The confinement of a water molecule between the cation and the anion in ImILs was demonstrated through NOESY experiment by our research group. This water can react reversibly with CO_2 to yield bicarbonate. The phenomenon occurred only with basic anions, such as $[OAc]^-$ and $[Im]^-$. Whereas fluorinated anions, such as $[BF_4]^-$ and $[NTf_2]^-$, show only noncovalent interactions with CO_2, and no significant contacts between water and cation could be detected. In the same work, ^{13}C inverse-gated decoupled analysis was employed to quantify the CO_2 capture by an aqueous solution of $[C_4C_1C_1Im]$[Im] and $[C_4C_1C_1Im]$[OAc]. The best system $[C_4C_1C_1Im]$[Im]/H_2O (1/1000) presented a sorption capacity of 10 molCO_2/molIL. Additionally, the comparison between the ^1H,^{15}N-HMBC NMR spectra of $[C_4C_1C_1Im]$[Im]/D_2O (1/1000) neat and under 20 bar of CO_2 reinforced the idea of bicarbonate formation instead of carbamate, since it was not observed any change in the chemical shift [48].

The formation of bicarbonate was demonstrated, once again, by Dupont and co-workers for bare ImILs under 10 bar of CO_2. In this work, the water in the

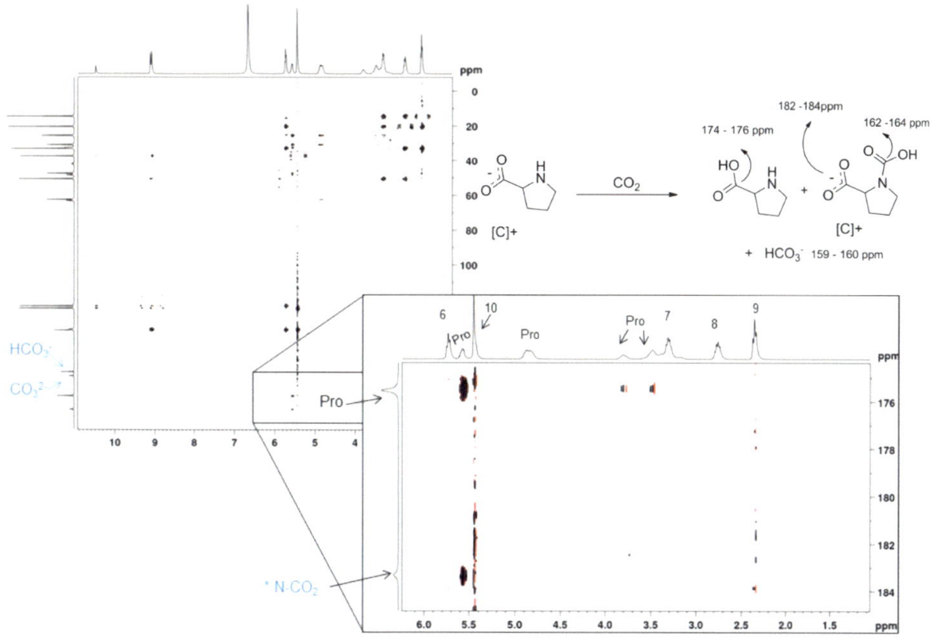

Figure 7.
Possible reaction of [C_4C_1Im][Pro] after bubbling CO_2 and $^1H,^{13}C$-HMBC NMR experiment (100 MHz, DMSO-d_6 capillary) (adapted from Ref. [49]).

hygroscopic IL was enough for the signal related to HCO_3^-/CO_3^{2-} species (ca. 160 ppm) to appear in ^{13}C NMR. A description of interactions between ImIL and CO_2 was demonstrated combining NMR spectroscopy with MD simulation. According to the ImIL structure, three types of CO_2-IL interaction were described: physical sorption by the polar domains (halogenated anions); chemical sorption by reaction with cation and formation of [C_4C_1Im]-CO_2 complex or by reaction with anion by carbamate formation (amino acid anions); and bicarbonate formation (hygroscopic basic anions). Another interesting result to point out in this work is [C_4C_1Im][Pro]-CO_2 interaction. Through the contour map of $^1H,^{13}C$-HMBC associated with ^{13}C NMR studies, it was possible to understand the CO_2 sorption mechanism and observe the pH dependence for this anion. Bubbling CO_2 in neat IL, the ^{13}C NMR signals suggested a reaction with the [Pro]$^-$ anion and the formation of the carbamate (ca. 180 and 162 ppm) and bicarbonate (ca. 161 ppm), which was confirmed by $^1H,^{13}C$-HMBC (**Figure 7**). This work demonstrated that the sorption capacities of ImILs could be enhanced by the variation of cation-anion-CO_2 and IL-CO_2-water interaction, providing a helpful investigation to understand the structure-activity relation in IL materials for CO_2 capture [49].

4. Interactions between ionic liquids and cellulose

Cellulose is a natural, renewable, and the most abundant polymer in nature [50]. The applications range from the papermaking industry and film manufacturing to pharmaceuticals. Cellulose's poor solubility in common solvents is attributed to inter and intramolecular H-bonds, which limits its uses and imposes very harsh treatments like the Lyocell process (N-methylmorpholine N-oxide dissolution) and solvent systems such as LiOH/urea, dimethyl sulfoxide (DMSO)/[$N_{4,4,4,4}$]F, and dimethylacetamide (DMAc)/LiCl.

ILs have been used successfully to dissolve cellulose either using pure ILs such as $[C_4C_1Im]Cl$ and $[C_4C_1Im][OAc]$ or using binary mixtures with organic solvents such as DMSO.

Although there are several strategies to dissolve this polymer, the dissolution process is still poorly understood. Alkali salts like LiOH or NaOH mixtures with urea are one of the most employed solvent systems. The dissolution and gelation of cellulose in 8 wt % $NaOH/H_2O$ with and without the addition of urea were studied using polarization transfer NMR techniques (PTNMR), INEPT, and CP [51]. It was revealed that the solvent containing urea is able to fully dissolve cellulose. The use of liquid state polarization technique (INEPT) and a solid-state polarization technique (CP) allowed the discrimination of liquid and solid states. It was noted by the authors that the presence of some signal in the fully dissolved urea/$NaOH/H_2O$ solution is due to the reorientation dynamics of a very viscous sample, leading to some CP transfer.

The dissolution of cellulose in LiOH/urea was studied using solid-state NMR (ssNMR), by using ^{13}C CP/MAS NMR, ^{6}Li CP/MAS NMR, ^{6}Li single-pulse NMR, 2D heteronuclear ^{1}H,^{13}C, and ^{1}H,^{6}Li-(FSLG-HETCOR) [52]. Four existing forms of Li^+ ions were detected. Some Li^+ ions coordinate with cellulose chain, and a large population of Li^+ ions in the form of Li^+-OH^--urea surrounds the Li^+-cellulose chains, which prevent their self-aggregation. The ability to dissolve cellulose is likely determined by the coordination ability of alkali metal ions (Li^+, Na^+, K^+).

XRD and NMR methods were used to study the low-temperature dissolving mechanism of chitin/chitosan in the alkali (LiOH, NaOH, and KOH) aqueous solvents [53]. Aqueous alkali showed different dissolving abilities based on the acetylation degree (AD) of chitin, namely, KOH > NaOH >> LiOH for AD between 0.94 and 0.74; KOH ≈ LiOH ≈ NaOH for AD between 0.53 and 0.25; and the inverse order, LiOH > > KOH > NaOH for AD lower than 0.25. By measuring the active species of cations in semi-frozen solutions using ^{7}Li, ^{23}N, and ^{39}K NMR, differences from the stoichiometric quantities were obtained with a Na^+/Li^+ ratio higher than K^+/Li^+, which is the proposed explanation to the sodium higher destabilizing effect of chitosan at lower temperatures.

$[C_nC_1Im][(C_1O)PHO_2]$ was used to dissolve and regenerate cellulose under mild conditions [54]. Requiring heating to 80°C and being able to dissolve up to 4 wt%, coagulation with water afforded cellulose as a transparent gel. NMR was used to monitor the dissolution process by measuring ^{31}P and ^{13}C chemical shift differences to the pure IL. PLM photographs show complete dissolution after 120 min of heating.

Minnick et al. studied the solubility of cellulose in mixtures of $[C_2C_1Im][(C_2O)_2PO_2]$ with aprotic solvents (DMSO, dimethylformamide, 1,3-dimethyl-2-imidazolidinone) and protic antisolvents (water, methanol, and ethanol), concluding that polar aprotic solvents/IL mixtures can improve the cellulose solubility, while polar solvents have the opposite effect [55]. The ^{1}H $\Delta\delta$ of the IL-cation imidazolium ring resonances presented downfield shifts for the aprotic solvents, while protic solvents have the opposite effect. The reduced shielding experienced by the IL in aprotic solvents can be attributed to the solvation of the cation, which decreases cation-anion interactions, resulting in an increase in anion availability to form more and stronger interactions with cellulose. The authors explained negative chemical shift variations of IL in protic solvents as a result of stronger IL-solvent interactions, which change/compete with the IL-cellulose interactions.

On the use of an organic solvent as co-solvent of cellulose, DMSO was used in conjunction with $[C_nC_1Im]Cl$ to dissolve cellulose, using NMR relaxation measurements to study the dissolution process and co-solvation ability of DMSO [56]. At low concentration, DMSO improves the cellulose solvating ability of $[C_nC_1Im]Cl$ by reducing the viscosity of the IL but weakens it at high concentration.

T_2 measurements on ^1H and 1D spectra on ^1H, ^{35}Cl indicate that the tight association between the cation and the anion in the $[C_nC_1Im]Cl$ network is loosened at low DMSO concentration. Neat $[C_nC_1Im]Cl$ exhibits one population with smaller T_2 having a slower rotational dynamics and another with a larger T_2 and faster rotational dynamics. When DMSO concentration was increased, the $[C_nC_1Im]Cl$ with slower rotational dynamics transfers to faster rotational dynamics by going through a coalescence step at 0.4 DMSO molar fraction. Analysis of $\Delta\delta$ allowed probing the ion clustering, by studying the imidazolium H-bond change with DMSO addition and by using ^{35}Cl NMR to access the chlorine local environment. At 0.5 DMSO molar fraction, a turning point in ^{35}Cl NMR shifts is also observed, in a similar way as in T_2 relaxation. As imidazolium ^1H shifts change due to the shielding effect, the higher concentration of DMSO breaks the IL H-bond network into small ion clusters that are incapable of dissolving cellulose.

NMR PFG-diffusion was used to study the dissolving process of microcrystalline cellulose in aqueous tetrabutylammonium hydroxide $[N_{4,4,4,4}][OH]$ at 40 wt% concentration [57]. From the molecular self-diffusion coefficients of microcrystalline cellulose (MCC) and $[N_{4,4,4,4}][OH]$, $[N_{4,4,4,4}]^+$ is involved in the dissolution of carbohydrates, binding in a 1.2 stoichiometry to glucose units based on diffusion measurements and mass relations. The binding might be due to electrostatic interactions between the deprotonated hydroxyl groups or to hydrophobic effects, which have been discussed in the literature as important in cellulose dissolution by using either NMR or other techniques, such as calorimetry and rheology [58, 59].

Another use of cellulose is to make physical composites by blending it with another polymer to reinforce the final material. To understand the molecular origin of the enhanced mechanical properties for a cellulose/Silk blend film (SF), ssNMR was used as the tool to investigate the conformational changes, intermolecular interactions between cellulose and silk, and the water organization [60]. ssNMR can be used to gain an insight into polymer's molecular structure and dynamics, being able to identify its molecular constituents, structural morphology, and chain dynamics [61].

Cellulose/silk films were prepared by dissolution in $[C_4C_1Im]Cl$ followed by precipitation with methanol. 2D ^1H,^{13}C-HETCOR NMR reveals the intermolecular H-bonding interactions between the NH groups of SF and the OH groups bonded to carbons C_2 and C_3 of cellulose. 2D WISE experiments measured the hydrophilic character of each carbon sites, based on the lineshape with increasing mixing times, water is more localized in the vicinity of cellulose chains than silk.

5. Ionic liquids as co-solvents

ILs can be used as co-solvents in several applications, not only to increase the solubility of target species but also to tune the important physicochemical properties, such as conductivity and viscosity toward a more efficient industrial performance and operational conditions. In recent years, mixtures of ILs either with other ILs (IL/IL mixtures) [62, 63], molecular liquids (water, organic solvents) [64–66], or polymers [67] have gained such an interest that has been labeled as the fourth evolution of ILs [68]. IL mixtures do not fall in the definition of salt in solution, neither their physicochemical properties follow mixing behavior of molecular liquid mixtures, which create the need to develop new fundamental and structural/dynamic research.

In recent years, there has been an interest on mixtures of two or more ILs (IL/IL mixtures). This approach has been presented as a new strategy to tune the IL properties, with several studies showing that in most mixtures, physicochemical properties have close to ideal mixing behavior, due to a random ionic distribution (Coulombic interactions). However, non-ideal behavior has been reported, particularly for mixtures with very different constituents [69, 70]. These "anomalies" can be explained by the formation of nano-segregated phases due to strong ionic interactions [71], making the ionic distribution nonrandom. Therefore, in terms of nanostructure organization, IL mixtures show an increased degree of complexity compared to neat ILs, having to consider the different combinations of the several ions present in solution. NMR techniques have been particularly useful to enlighten these deviations from linearity. Matthews et al. analyzed binary mixtures of ILs with a common cation $[C_4C_1Im]^+$ and different anions by ^{1}H and ^{13}C NMR chemical shift peak analysis [72]. Studying the full concentration range of IL/IL mixtures, they inferred that there is a preferential interaction of the cation with one of the anions. That was evident by the H_2 ^{1}H NMR peak, which shows a preferential interaction with the stronger H-bond acceptor anion over the NMR timescale. From the ^{13}C NMR, imidazolium ring C_4 and C_5 showed greater affinity for the weakest H-bond anions, which was not obvious in ^{1}H NMR, suggesting that the weakest H-bond will occupy a position above the plane of the imidazolium, forming an anion-π interaction. These slightly preferential associations have a very small effect on the structural organization but can be the driving force for nonideal deviations observed for some ILs.

Marullo et al. used $[BzC_4Im][NTf_2]$ with $[Bz(F_5)C_4Im][NTf_2]$ or $[BzC_4Im][BF_4]$ or $[N_{3,3,3,4}][NTf_2]$, as a function of the composition, as solvents for the Diels Alder reaction of 9-anthracenemethanol with N-ethylmaleimide [73]. The authors observed that the reactivity is affected by two main solvent parameters: the viscosity and the molecular organization. For IL/IL mixtures sharing the same cation, the kinetic rate was showed to be more affected by composition, correlating to structural modifications due to a more ordered local domain. For the anion shared IL/IL mixtures, they observed an unexpected result: changing the molar ratio of two cations with similar size and chemical nature produced a more profound effect on the reaction kinetics than IL mixtures $([BzC_4Im]_x[N_{3,3,3,4}]_{1-x}[NTf_2])$ with different size cations. ^{1}H Δδ showed that the imidazolium H_2, H_9, and H_{10} exhibit a significant upfield shift, in which H_2 and H_9 show two chemical shift domains, with a discontinuity point, $[BzC_4Im]_{0.4}[N_{3,3,3,4}]_{0.6}[NTf_2]$, corresponds to the maximum rate, to which the authors justified with the transition point between a typical aliphatic IL and an aromatic type IL.

One of the applications of ILs is in the field of tribology [74]. ILs can be used as a lubricant by itself or as an additive to increase commercially available oil efficiency. Combining low molecular weight polyethylene glycol (PEG) with orthoborate ILs allowed a decrease in the wear and friction properties for steel on steel contacts. Filippov et al. studied mixtures of PEG 200 with ImILs derived from $[BMB]^-$ or $[BOB]^-$ anions as function of concentration by diffusion NMR and Δδ [75]. The chemical shift analysis showed that the presence of PEG affects the chemical shift of both the cations and the anions, with the protons of the imidazolium ring of the cation the most affected. For the anions, $[BMB]^-$ showed two chemical shift domains in the PEG concentration range, suggesting that above 50% wt PEG, there is a modification on the interaction mode of PEG/IL, which is not visible for the ImIL derived from $[BOB]^-$. The diffusion experiments showed that for ImIL derived from $[BMB]^-$, the cation and anion differ on their diffusion coefficient in all the concentration range, which suggested that in the mixture, the IL is partially dissociated, which agrees with the data obtained from impedance analysis.

6. Conclusions

The combination of different NMR experiments allowed to observe the molecular panorama of cation-anion-solute interactions. The diversity of NMR techniques at our disposal allows gathering crucial information to rationalize the IL solvent behavior. Despite the notion that ILs are tailored solvents, their optimization is still seldom done on the basis of the fundamental knowledge of IL performance and the dissolution mechanisms. Instead, the trial and error approach often prevails.

The success of ILs as solvents depends on the strength and nature of the interactions that cations and anions can establish with the solute while maintaining their intrinsic interaction. Chemical shift deviation analyses identify the electronic perturbations in nuclei, either by making or breaking H-bonds or by inducing changes in the geometry. NOE experiments allow a more complete picture, probing solute structural organization within ILs; in this case, both strong and weak interactions can be identified. Relaxation studies are capable of sensing a shorter timescale and identifying even short-lived events that take place upon solubilization processes. Finally, diffusion studies in ILs provide the knowledge of solute behavior and respective associations and also cation/anion feedback in the presence of the solute. This information allows probing macroscopic properties, such as viscosity, which are intimately related to the desired application of ILs.

Acknowledgements

This work was funded by FCT—Portuguese Foundation for Science and Technology, Portugal, under the projects PTDC/EPH-PAT/0224/2014, PTDC/QUI-QFI/31508/2017, POR Lisboa, and PTNMR (ROTEIRO/0031/ 2013; PINFRA/22161/2016), co-financed by FEDER through COMPETE 2020, Portugal, POCI, and PORL and FCT through PIDDAC (POCI-01-0145-FEDER-007688; UID/CTM/50025/2019; and UID/Multi/04378/2019). T.G.P. acknowledges FCT for the SFRH/BD/133447/2017 PhD fellowship.

Conflict of interest

The authors declare no conflict of interest.

Abbreviations

Cations

$[(i\mathrm{Pr})C_2C_1\mathrm{Im}]^+$	1-methyl-3-isopentylimidazolium
$[a C_2C_1C_1\mathrm{Im}]^+$	1,2-dimethyl-(3-aminoethyl)imidazolium
$[\mathrm{Bz}(F_5)C_4\mathrm{Im}]^+$	1-(2,3,4,5,6-pentafluorophenyl)-3-butylimidazolium
$[\mathrm{Bz}C_1\mathrm{Im}]^+$	1-benzyl-3-methylimidazolium
$[\mathrm{Bz}C_4\mathrm{Im}]^+$	1-benzyl-3-butylimidazolium
$[C_1C_1\mathrm{Im}]^+$	1,3-dimethylimidazolium
$[C_2C_1\mathrm{Im}]^+$	1-ethyl-3-methylimidazolium
$[C_4C_1C_1\mathrm{Im}]^+$	1-butyl-2,3-dimethylimidazolium
$[C_4C_1\mathrm{Im}]^+$	1-butyl-3-methylimidazolium
$[C_4C_1\mathrm{Pyr}]^+$	1-butyl-1-methylpyrrolidinium
$[C_4\mathrm{Py}]^+$	1-butylpyridinium
$[C_nC_1\mathrm{Im}]^+$	1-alkyl-3-methylimidazolium
$[C_nC_1\mathrm{Pip}]^+$	1-butyl-1-methylpiperidinium
$[C_nC_1\mathrm{Pyr}]^+$, n = 3, 4, 6, 8, and 10	1-alkyl-1-methylpyrrolidinium
$[N_{1,1,1,1}]^+$	tetramethylammonium

$[N_{1,1,1,H}]^+$	trimethylammonium
$[N_{1,1,n,2OH}]^+$	alkyldimethyl(2-hydroxyethyl)ammonium
$[N_{3,3,3,4}]^+$	n-butyltriethylammonium
$[P_{4,4,4,2}]^+$	tri-n-butylethylphosphonium
$[P_{6,6,6,14}]^+$	trihexyl(tetradecyl)phosphonium

Anions

$[BF_4]^-$	tetrafluoroborate
$[BMB]^-$	bis(mandelato)borate
$[BOB]^-$	bis(oxalato)borate
$[2\text{-}CNPyr]^-$	2-cianopyrrolide
$[(C_1O)PHO_2]^-$	methylphosphonate
$[C_1SO_3]^-$	methylsulfonate
$[C_1SO_4]^-$	methylsulfate
$[C_2SO_4]^-$	ethylsulfate
$[DCA]^-$	dicyanamide
$[(C_2O)_2PO_2]^-$	diethyl phosphate
$[Im]^-$	imidazolate
$[Lys]^-$	lysinate
$[NO_3]^-$	nitrate
$[NTf_2]^-$	bis(trifluoromethylsulfonyl)imide
$[OAc]^-$	acetate
$[OTf]^-$	trifluoromethanesulfonate
$[PF_6]^-$	hexafluorophosphate
$[Pro]^-$	prolinate
$[SCN]^-$	thiocyanate
$[Suc]^-$	succinate
$[Tau]^-$	taurine
$[TCM]^-$	tricyanomethanide
$[Threo]^-$	threonine

Techniques

COSY	$^1H,^1H$-correlation spectroscopy
CP	cross polarization
DFT	density functional theory
DOSY	diffusion ordered spectroscopy
FSLG-HETCOR	frequency-switched Lee-Goldburg—dipolar heteronuclear correlation
HMBC	heteronuclear multiple bond correlation
HSQC	heteronuclear single quantum coherence
HOESY	heteronuclear nuclear Overhauser effect spectroscopy
INADEQUATE	incredible natural abundance double quantum transfer experiment
INEPT	insensitive nuclei enhanced by polarization transfer
IR	infrared
FT	Fourier-transform
MAS	magic angle spinning
MD	molecular dynamic
NMR	nuclear magnetic resonance
NOE	nuclear Overhauser effect
NOESY	nuclear Overhauser effect spectroscopy

PFG	pulsed field gradient
PLM	polarized light microscopy
PTNMR	polarization transfer NMR
ssNMR	solid state nuclear magnetic resonance
XRD	X-ray diffraction

Author details

Mónica M. Lopes[1], Raquel V. Barrulas[1], Tiago G. Paiva[1], Ana S.D. Ferreira[2], Marcileia Zanatta[1] and Marta C. Corvo[1*]

1 School of Science and Technology, Materials Science Department, i3N|Cenimat, NOVA University Lisbon, Caparica, Portugal

2 School of Science and Technology, Chemistry Department, UCIBIO, NOVA University Lisbon, Caparica, Portugal

*Address all correspondence to: marta.corvo@fct.unl.pt

References

[1] Hayes R, Warr GG, Atkin R. Structure and nanostructure in ionic liquids. Chemical Reviews. 2015;**115**:6357-6426. DOI: 10.1021/cr500411q

[2] Damodaran K. Recent NMR studies of ionic liquids. In: Annual Reports on NMR Spectroscopy. Elsevier. 2016. Vol. 88 pp. 215-244. DOI: 10.1016/bs.arnmr.2015.11.002

[3] Nanda R, Damodaran K. A review of NMR methods used in the study of the structure and dynamics of ionic liquids. Magnetic Resonance in Chemistry. 2018;**56**:62-72. DOI: 10.1002/mrc.4666

[4] Zheng YZ, Wang NN, Luo JJ, Zhou Y, Yu ZW. Hydrogen-bonding interactions between [BMIM][BF4] and acetonitrile. Physical Chemistry Chemical Physics. 2013;**15**:18055-18064. DOI: 10.1039/c3cp53356e

[5] Wang H, Cui J, Li H, Zhao Y, Wang J. The effect of cationic structure of ionic liquids on their interactions with 5-hydroxymethylfurfural. Journal of Molecular Structure. 2019;**1179**:57-64. DOI: 10.1016/j.molstruc.2018.10.093

[6] Khatun S, Castner EW. Ionic liquid-solute interactions studied by 2D NOE NMR spectroscopy. The Journal of Physical Chemistry. B. 2015;**119**:9225-9235. DOI: 10.1021/jp509861g

[7] Castiglione F, Appetecchi GB, Passerini S, Panzeri W, Indelicato S, Mele A. Multiple points of view of heteronuclear NOE: Long range vs short range contacts in pyrrolidinium based ionic liquids in the presence of Li salts. Journal of Molecular Liquids. 2015;**210**:215-222. DOI: 10.1016/j.molliq.2015.05.036

[8] Castiglione F, Famulari A, Raos G, Meille SV, Mele A, Appetecchi GB, et al. Pyrrolidinium-based ionic liquids doped with lithium salts: How does Li+ coordination affect its diffusivity? The Journal of Physical Chemistry. B. 2014;**118**:13679-13688. DOI: 10.1021/jp509387r

[9] Sebastião RCO, Pacheco CN, Braga JP, Piló-Veloso D. Diffusion coefficient distribution from NMR-DOSY experiments using Hopfield neural network. Journal of Magnetic Resonance. 2006;**182**:22-28. DOI: 10.1016/j.jmr.2006.06.005

[10] Kaintz A, Baker G, Benesi A, Maroncelli M. Solute diffusion in ionic liquids, NMR measurements and comparisons to conventional solvents. The Journal of Physical Chemistry. B. 2013;**117**:11697-11708. DOI: 10.1021/jp405393d

[11] Garaga MN, Nayeri M, Martinelli A. Effect of the alkyl chain length in 1-alkyl-3-methylimidazolium ionic liquids on inter-molecular interactions and rotational dynamics. Journal of Molecular Liquids. 2015;**210**:169-177. DOI: 10.1016/j.molliq.2015.06.055

[12] Rumble CA, Kaintz A, Yadav SK, Conway B, Araque JC, Baker GA, et al. Rotational dynamics in ionic liquids from NMR relaxation experiments and simulations: Benzene and 1-ethyl-3-methylimidazolium. The Journal of Physical Chemistry. B. 2016;**120**:9450-9467. DOI: 10.1021/acs.jpcb.6b06715

[13] Rumble CA, Uitvlugt C, Conway B, Maroncelli M. Solute rotation in ionic liquids: Size, shape, and electrostatic effects. The Journal of Physical Chemistry. B. 2017;**121**:5094-5109. DOI: 10.1021/acs.jpcb.7b01704

[14] Potdar S, Anantharaj R, Banerjee T. Aromatic extraction using mixed ionic liquids: Experiments and COSMO-RS predictions. Journal of Chemical and

Engineering Data. 2012;**57**:1026-1035. DOI: 10.1021/je200924e

[15] Revelli AL, Mutelet F, Jaubert JN. Extraction of benzene or thiophene from n-heptane using ionic liquids. NMR and thermodynamic study. The Journal of Physical Chemistry. B. 2010;**114**:4600-4608. DOI: 10.1021/jp911978a

[16] Kumar AAP, Banerjee T. Thiophene separation with ionic liquids for desulphurization: A quantum chemical approach. Fluid Phase Equilibria. 2009;**278**:1-8. DOI: 10.1016/j.fluid.2008.11.019

[17] Dias N, Shimizu K, Morgado P, Filipe EJM, Canongia Lopes JN, Vaca Chávez F. Charge templates in aromatic plus ionic liquid systems revisited: NMR experiments and molecular dynamics simulations. The Journal of Physical Chemistry. B. 2014;**118**:5772-5780. DOI: 10.1021/jp503130y

[18] Shimizu K, Gomes MFC, Padua AAH, Rebelo LPN, Lopes JNC. On the role of the dipole and quadrupole moments of aromatic compounds in the solvation by ionic liquids. The Journal of Physical Chemistry. B. 2009;**113**: 9894-9900. DOI: 10.1021/jp903556q

[19] Su B-MM, Zhang S, Zhang ZC. Structural elucidation of thiophene interaction with ionic liquids by multinuclear NMR spectroscopy. The Journal of Physical Chemistry. B. 2004;**108**:19510-19517. DOI: 10.1021/jp049027l

[20] Barrulas RV, Paiva TG, Corvo MC. NMR methodology for a rational selection of ionic liquids: Extracting polyphenols. Separation and Purification Technology. 2019;**221**:29-37. DOI: 10.1016/j.seppur.2019.03.077

[21] Román-Leshkov Y, Barrett CJ, Liu ZY, Dumesic JA. Production of dimethylfuran for liquid fuels from biomass-derived carbohydrates. Nature. 2007;**447**:982-985. DOI: 10.1038/nature05923

[22] Wang NN, Zhang QG, Wu FG, Li QZ, Yu ZW. Hydrogen bonding interactions between a representative pyridinium-based ionic liquid [BuPy][BF4] and water/dimethyl sulfoxide. The Journal of Physical Chemistry. B. 2010;**114**: 8689-8700. DOI: 10.1021/jp103438q

[23] Perinu C, Arstad B, Jens K-J. NMR spectroscopy applied to amine–CO_2–H_2O systems relevant for post-combustion CO_2 capture: A review. International Journal of Greenhouse Gas Control. 2014;**20**:230-243. DOI: 10.1016/J.IJGGC.2013.10.029

[24] Corvo MC, Sardinha J, Menezes SC, Einloft S, Seferin M, Dupont J, et al. Solvation of carbon dioxide in [c4mim][bf4] and [c4mim][pf6] ionic liquids revealed by high-pressure NMR spectroscopy. Angewandte Chemie, International Edition. 2013;**52**: 13024-13027. DOI: 10.1002/anie.201305630

[25] Corvo MC, Sardinha J, Casimiro T, Marin G, Seferin M, Einloft S, et al. A rational approach to CO_2 capture by imidazolium ionic liquids: Tuning CO_2 solubility by cation alkyl branching. ChemSusChem. 2015;**8**:1935-1946. DOI: 10.1002/cssc.201500104

[26] Allen J, Damodaran K. High-resolution slice selection NMR for the measurement of CO_2 diffusion under non-equilibrium conditions. Magnetic Resonance in Chemistry. 2015;**53**: 200-202. DOI: 10.1002/mrc.4176

[27] Blath J, Deubler N, Hirth T, Schiestel T. Chemisorption of carbon dioxide in imidazolium based ionic liquids with carboxylic anions. Chemical Engineering Journal. 2012;**181-182**: 152-158. DOI: 10.1016/j.cej.2011.11.042

[28] Besnard M, Cabaço MI, Vaca Chávez F, Pinaud N, Sebastião PJ,

Coutinho JAP, et al. CO_2 in 1-butyl-3-methylimidazolium acetate. 2. NMR investigation of chemical reactions. The Journal of Physical Chemistry. A. 2012;**116**:4890-4901. DOI: 10.1021/jp211689z

[29] Cabaço MI, Besnard M, Chávez FV, Pinaud N, Sebastião PJ, Coutinho JAP, et al. Understanding chemical reactions of CO_2 and its isoelectronic molecules with 1-butyl-3-methylimidazolium acetate by changing the nature of the cation: The case of CS_2 in 1-butyl-1-methylpyrrolidinium acetate studied by NMR spectroscopy and density functional theory calculations. The Journal of Chemical Physics. 2014;**140**:244307. DOI: 10.1063/1.4884820

[30] Kortunov PV, Baugh LS, Siskin M. Pathways of the chemical reaction of carbon dioxide with ionic liquids and amines in ionic liquid solution. Energy and Fuels. 2015;**29**:5990-6007. DOI: 10.1021/acs.energyfuels.5b00876

[31] Umecky T, Abe M, Takamuku T, Makino T, Kanakubo M. CO_2 absorption features of 1-ethyl-3-methylimidazolium ionic liquids with 2,4-pentanedionate and its fluorine derivatives. Journal of CO_2 Utilization. 2019;**31**:75-84. DOI: 10.1016/j.jcou.2019.02.020

[32] Seo S, Quiroz-Guzman M, Gohndrone TR, Brennecke JF. Comment on Selective chemical separation of carbon dioxide by ether functionalized imidazolium cation based ionic liquids. In: Sharma P, Choi SH, Park Sd, Baek IH, and Lee GS, editors. The Chemical Engineering Journal. 2012;**181-182**: 834-841. The Chemical Engineering Journal. 2014;**245**:367-369. DOI:10.1016/j.cej.2013.08.051

[33] Sharma P, Choi SH, Do Park S, Baek IH, Lee GS. Selective chemical separation of carbondioxide by ether functionalized imidazolium cation based ionic liquids. Chemical Engineering Journal. 2012;**181-182**: 834-841. DOI: 10.1016/j.cej.2011.12.024

[34] Bates ED, Mayton RD, Ntai I, Davis JH. CO_2 capture by a task-specific ionic liquid. Journal of the American Chemical Society. 2002;**124**:926-927. DOI: 10.1021/ja017593d

[35] Kortunov PV, Siskin M, Baugh LS, Calabro DC. In situ nuclear magnetic resonance mechanistic studies of carbon dioxide reactions with liquid amines in non-aqueous systems: Evidence for the formation of carbamic acids and zwitterionic species. Energy and Fuels. 2015;**29**:5940-5966. DOI: 10.1021/acs.energyfuels.5b00985

[36] Kortunov PV, Siskin M, Baugh LS, Calabro DC. In situ nuclear magnetic resonance mechanistic studies of carbon dioxide reactions with liquid amines in aqueous systems: New insights on carbon capture reaction pathways. Energy and Fuels. 2015;**29**:5919-5939. DOI: 10.1021/acs.energyfuels.5b00850

[37] Bhattacharyya S, Shah FU. Ether functionalized choline tethered amino acid ionic liquids for enhanced CO_2 capture. ACS Sustainable Chemistry and Engineering. 2016;**4**:5441-5449. DOI: 10.1021/acssuschemeng.6b00824

[38] Bhattacharyya S, Filippov A, Shah FU. Insights into the effect of CO_2 absorption on the ionic mobility of ionic liquids. Physical Chemistry Chemical Physics. 2016;**18**:28617-28625. DOI: 10.1039/C6CP05804C

[39] Chen Y, Zhou XQ, Cao Y, Xue Z, Mu T. Quantitative investigation on the physical and chemical interactions between CO_2 and amine-functionalized ionic liquid [aEMMIM][BF 4] by NMR. Chemical Physics Letters. 2013;**574**:124-128. DOI: 10.1016/j.cplett.2013.04.069

[40] Filippov A, Bhattacharyya S, Shah FU. CO_2 absorption and ion mobility in aqueous choline-based ionic liquids. Journal of Molecular Liquids. 2019;**276**:748-752. DOI: 10.1016/J.MOLLIQ.2018.12.045

[41] Goodrich BF, de la Fuente JC, Gurkan BE, Lopez ZK, Price EA, Huang Y, et al. Effect of water and temperature on absorption of CO_2 by amine-functionalized anion-tethered ionic liquids. The Journal of Physical Chemistry. B. 2011;**115**:9140-9150. DOI: 10.1021/jp2015534

[42] Seo S, Quiroz-Guzman M, Desilva MA, Lee TB, Huang Y, Goodrich BF, et al. Chemically tunable ionic liquids with aprotic heterocyclic anion (AHA) for CO_2 capture. The Journal of Physical Chemistry. B. 2014;**118**:5740-5751. DOI: 10.1021/jp502279w

[43] Huang Y, Cui G, Zhao Y, Wang H, Li Z, Dai S, et al. Preorganization and cooperation for highly efficient and reversible capture of low-concentration CO_2 by ionic liquids. Angewandte Chemie, International Edition. 2017;**56**:13293-13297. DOI: 10.1002/anie.201706280

[44] Huang Y, Cui G, Wang H, Li Z, Wang J. Tuning ionic liquids with imide-based anions for highly efficient CO_2 capture through enhanced cooperations. Journal of CO_2 Utilization. 2018;**28**: 299-305. DOI: 10.1016/J.JCOU.2018.10.013

[45] Zanatta M, Simon NM, dos Santos FP, Corvo MC, Cabrita EJ, Dupont J. Correspondence on "preorganization and cooperation for highly efficient and reversible capture of low-concentration CO_2 by ionic liquids". Angewandte Chemie, International Edition. 2019;**58**:382-385. DOI: 10.1002/anie.201712252

[46] McDonald JL, Sykora RE, Hixon P, Mirjafari A, Davis JH. Impact of water on CO_2 capture by amino acid ionic liquids. Environmental Chemistry Letters. 2014;**12**:201-208. DOI: 10.1007/s10311-013-0435-1

[47] Filippov A, Antzutkin ON, Shah FU. Reactivity of CO_2 with aqueous choline-based ionic liquids probed by solid-state NMR spectroscopy. Journal of Molecular Liquids. 2019;**286**:110918. DOI: 10.1016/J.MOLLIQ.2019.110918

[48] Simon NM, Zanatta M, dos Santos FP, Corvo MC, Cabrita EJ, Dupont J. Carbon dioxide capture by aqueous ionic liquid solutions. ChemSusChem. 2017;**10**:4927-4933. DOI: 10.1002/cssc.201701044

[49] Simon NM, Zanatta M, Neumann J, Girard A-L, Marin G, Stassen H, et al. Cation–anion–CO_2 interactions in imidazolium-based ionic liquid sorbents. ChemPhysChem. 2018;**19**:2879-2884. DOI: 10.1002/cphc.201800751

[50] Kafy A, Ko H-U, Kim HC, Mun S, Zhai L, Kim J. Renewable smart materials. Smart Materials and Structures. 2016;**25**:073001. DOI: 10.1088/0964-1726/25/7/073001

[51] Alves L, Medronho B, Filipe A, E. Antunes F, Lindman B, Topgaard D, et al. New insights on the role of urea on the dissolution and thermally-induced gelation of cellulose in aqueous alkali. Gels. 2018;**4**:87. DOI: 10.3390/gels4040087

[52] Huang H, Ge H, Song J, Yao Y, Chen Q, Xu M. NMR study on the roles of Li+ in the cellulose dissolution process. ACS Sustainable Chemistry and Engineering. 2019;**7**:618-624. DOI: 10.1021/acssuschemeng.8b04177

[53] Ru G, Wu S, Yan X, Liu B, Gong P, Wang L, et al. Inverse solubility of chitin/

chitosan in aqueous alkali solvents at low temperature. Carbohydrate Polymers. 2019;**206**:487-492. DOI: 10.1016/j.carbpol.2018.11.016

[54] Xu K, Xiao Y, Cao Y, Peng S, Fan M, Wang K. Dissolution of cellulose in 1-allyl-3-methylimidazolium methyl phosphonate ionic liquid and its composite system with Na_2PHO_3. Carbohydrate Polymers. 2019;**209**: 382-388. DOI: 10.1016/j.carbpol.2018.12.040

[55] Minnick DL, Flores RA, DeStefano MR, Scurto AM. Cellulose solubility in ionic liquid mixtures: Temperature, cosolvent, and antisolvent effects. The Journal of Physical Chemistry. B. 2016;**120**:7906-7919. DOI: 10.1021/acs.jpcb.6b04309

[56] Zhang C, Kang H, Li P, Liu Z, Zhang Y, Liu R, et al. Dual effects of dimethylsulfoxide on cellulose solvating ability of 1-allyl-3-methylimidazolium chloride. Cellulose. 2016;**23**:1165-1175. DOI: 10.1007/s10570-016-0876-3

[57] Gentile L, Olsson U. Cellulose–solvent interactions from self-diffusion NMR. Cellulose. 2016;**23**:2753-2758. DOI: 10.1007/s10570-016-0984-0

[58] Paiva T, Echeverria C, Godinho MH, Almeida PL, Corvo MC. On the influence of imidazolium ionic liquids on cellulose derived polymers. European Polymer Journal. 2019;**114**:353-360. DOI: 10.1016/j.eurpolymj.2019.02.032

[59] Kostag M, El Seoud OA. Dependence of cellulose dissolution in quaternary ammonium-based ionic liquids/DMSO on the molecular structure of the electrolyte. Carbohydrate Polymers. 2019;**205**:524-532. DOI: 10.1016/j.carbpol.2018.10.055

[60] Tian D, Li T, Zhang R, Wu Q, Chen T, Sun P, et al. Conformations and intermolecular interactions in cellulose/silk fibroin blend films: A solid-state NMR perspective. The Journal of Physical Chemistry. B. 2017;**121**: 6108-6116. DOI: 10.1021/acs.jpcb.7b02838

[61] Saalwächter K, Spiess HW. Solid-state NMR of polymers. In: Polymer Science: A Comprehensive Reference. Vol. 10. Amsterdam: Elsevier; 2012. pp. 185-219. DOI: 10.1016/B978-0-444-53349-4.00025-X

[62] Niedermeyer H, Hallett JP, Villar-Garcia IJ, Hunt PA, Welton T. Mixtures of ionic liquids. Chemical Society Reviews. 2012;**41**:7780. DOI: 10.1039/c2cs35177c

[63] Chatel G, Pereira JFB, Debbeti V, Wang H, Rogers RD. Mixing ionic liquids—“Simple mixtures” or “double salts”? Green Chemistry. 2014;**16**:2051. DOI: 10.1039/c3gc41389f

[64] Lui MY, Crowhurst L, Hallett JP, Hunt PA, Niedermeyer H, Welton T. Salts dissolved in salts: Ionic liquid mixtures. Chemical Science. 2011;**2**:1491. DOI: 10.1039/c1sc00227a

[65] Iglesias-Otero MA, Troncoso J, Carballo E, Romaní L. Density and refractive index in mixtures of ionic liquids and organic solvents: Correlations and predictions. The Journal of Chemical Thermodynamics. 2008;**40**:949-956. DOI: 10.1016/j.jct.2008.01.023

[66] Reddy PN, Padmaja P, Subba Reddy BV, Rambabu G. Ionic liquid/water mixture promoted organic transformations. RSC Advances. 2015;**5**:51035-51054. DOI: 10.1039/C5RA08625F

[67] Rodríguez H, Rogers RD. Liquid mixtures of ionic liquids and polymers as solvent systems. Fluid Phase Equilibria. 2010;**294**:7-14. DOI: 10.1016/j.fluid.2009.12.036

[68] MacFarlane DR, Chong AL, Forsyth M, Kar M, Vijayaraghavan R, Somers A, et al. New dimensions in salt–solvent mixtures: A 4th evolution of ionic liquids. Faraday Discussions. 2018;**206**:9-28. DOI: 10.1039/C7FD00189D

[69] Canongia Lopes JN, Cordeiro TC, Esperança JMSS, Guedes HJR, Huq S, Rebelo LPN, et al. Deviations from ideality in mixtures of two ionic liquids containing a common ion. The Journal of Physical Chemistry. B. 2005;**109**: 3519-3525. DOI: 10.1021/jp0458699

[70] Omar S, Lemus J, Ruiz E, Ferro VR, Ortega J, Palomar J. Ionic liquid mixtures—An analysis of their mutual miscibility. The Journal of Physical Chemistry. B. 2014;**118**: 2442-2450. DOI: 10.1021/jp411527b

[71] Fillion JJ, Brennecke JF. Viscosity of ionic liquid–ionic liquid mixtures. Journal of Chemical and Engineering Data. 2017;**62**:1884-1901. DOI: 10.1021/acs.jced.7b00221

[72] Matthews RP, Villar-Garcia IJ, Weber CC, Griffith J, Cameron F, Hallett JP, et al. structural investigation of ionic liquid mixtures. Physical Chemistry Chemical Physics. 2016;**18**:8608-8624. DOI: 10.1039/C6CP00156D

[73] Marullo S, D'Anna F, Campodonico PR, Noto R. Ionic liquid binary mixtures: How different factors contribute to determine their effect on the reactivity. RSC Advances. 2016;**6**: 90165-90171. DOI: 10.1039/C6RA12836J

[74] Zhou Y, Qu J. Ionic liquids as lubricant additives: A review. ACS Applied Materials and Interfaces. 2017;**9**:3209-3222. DOI: 10.1021/acsami.6b12489

[75] Filippov A, Azancheev N, Gibaydullin A, Bhattacharyya S, Antzutkin ON, Shah FU. Dynamic properties of imidazolium orthoborate ionic liquids mixed with polyethylene glycol studied by NMR diffusometry and impedance spectroscopy. Magnetic Resonance in Chemistry. 2018;**56**: 113-119. DOI: 10.1002/mrc.4636

Chapter 5

New Advances in Fast Methods of 2D NMR Experiments

Abdul-Hamid Emwas, Mawadda Alghrably,
Samah Al-Harthi, Benjamin Gabriel Poulson,
Kacper Szczepski, Kousik Chandra and Mariusz Jaremko

Abstract

Although nuclear magnetic resonance spectroscopy is a potent analytical tool for identification, quantification, and structural elucidation, it suffers from inherently low sensitivity limitations. This chapter focuses on recently reported methods that enable quick acquisition of NMR spectra, as well as new methods of faster, efficient, and informative two-dimensional (2D) NMR methods. Fast and efficient data acquisition has risen in response to an increasing need to investigate chemical and biological processes in real time. Several new techniques have been successfully introduced. One example of this is band-selective optimized-flip-angle short-transient (SOFAST) NMR, which has opened the door to studying the kinetics of biological processes such as the phosphorylation of proteins. The fast recording of NMR spectra allows researchers to investigate time sensitive molecules that have limited stability under experimental conditions. The increasing awareness that molecular structures are dynamic, rather than static, has pushed some researchers to find alternatives to standard, time-consuming methods of ^{15}N relaxation observables acquisition.

Keywords: NMR, 2D NMR, ultrafast data processing, SOFAST, relaxation

1. Introduction

Nuclear magnetic resonance (NMR) spectroscopy is a powerful analytical tool for identifying, quantifying, and elucidating structures of molecules. Moreover, NMR offers several approaches to studying molecular dynamics and to probing the interactions of inert molecules such as protein-protein, protein-DNA, and protein-ligand. The main strength of NMR spectroscopy is its ability to discern or resolve individual resonance frequencies of a wide range of different nuclei (e.g., ^{1}H, ^{13}C, ^{15}N, ^{29}Si, ^{30}Al, ^{31}P to ^{235}U). As the resonating frequency is unique for each type of nuclei, one can envision an NMR for each nucleus as a separate spectroscopy such as ^{1}H NMR spectroscopy, ^{13}C NMR spectroscopy, ^{235}U NMR spectroscopy (**Figure 1**), etc. More importantly, NMR has the ability to evaluate information about the environment of each atom and their neighbor's nuclei (both through space and through bond), allowing researchers to differentiate the unique magnetic environments of the same nuclei in different positions of a single molecule. Thus, NMR spectroscopy is extensively used for the identification and in the structural elucidation in a wide

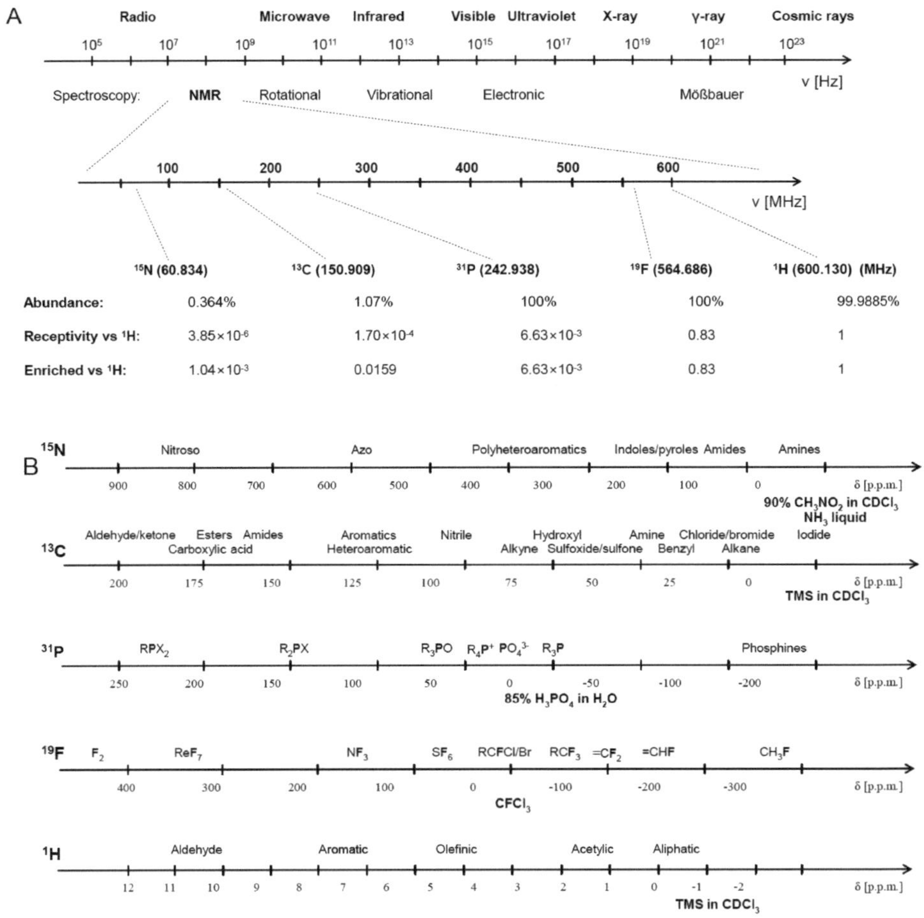

Figure 1.
Frequency scale ranges and types of spectroscopies that correspond to them, respectively. The natural abundances and chemical shifts of the most common (^{15}N, ^{13}C, ^{31}P, ^{19}F, and ^{1}H nuclei) nuclei biomolecules studies with respect to the scale of the 600 MHz proton frequency adapted from [1].

range of applications in gas [2, 3], liquid [4–16], and solid-state samples [17–28]. Nowadays, NMR spectroscopy is one of the most important analytical tools that has been used in several fields. These fields include structural biology [29–38], organic chemistry [39–52], polymer characterization [40, 46, 53–64], inorganic chemistry [65–75], and physics [76–83].

Despite its significant advantages, NMR suffers from some limitations, of which the relatively low sensitivity seems to be the most severe. An NMR sample can be treated as a collection of many nuclear spins of magnetically active nuclei that act as small bar magnets. These nuclear spins have two possible orientations with different energy levels that adapt when placed within the strong magnetic field. The number of nuclear spins occupying each energy level is determined by the Boltzmann distribution equation:

$$N_U/N_L = e^{-\Delta E/kT} = e^{-h\nu/kT}$$

where NU and NL are the nuclei (expressed in numbers) in the upper and lower energy states, respectively, k is the Boltzmann constant, ΔE is the energy gap between two energy states of the spins, higher (NU) and lower (NL) energy, respectively, and T is the temperature expressed in Kelvin (K). Many of the nuclei

DOI: http://dx.doi.org/10.5772/intechopen.90263

are at 298 K, the most common temperature in which biomolecular NMR measurements are performed. At this temperature, there is miniscule excess of the nuclei in the lower energy state that could be excited to the higher energy state by absorbing the energy given by the radio frequency (RF) pulse and later detected successfully during the data acquisition process. Therefore, the relatively low sensitivity of the NMR measurements seems to be the most severe limitation of this method. Another limitation comes from each measured nuclei having a defined range of frequencies at which it resonates (for practical reasons, frequencies are recalculated into the "unit-less" ppm scale in order to make it independent of the applied external magnetic field, B0). This in turn depends on its magnetic neighborhood, and, therefore, the whole range must be covered with good resolution.

In earlier days, most NMR experiments only recorded one-dimensional (1D) spectra of which several were of ^{1}H or ^{13}C nuclei. For complex biomacromolecules, such as proteins, nucleic acids and their complexes, and complex mixtures, this technique has a particularly low spectral range of ^{1}H resonance, what leads to the low resolution of the spectra. This low resolution proves insufficient for analysis.

NMR signals depend on several nuclear properties such as the natural abundance of investigated isotopes, nuclear spin (I), gyromagnetic ratio (γ), quadrupolar moment (Q), and spin relaxation rates. Although most elements have an NMR active isotope, NMR spectroscopy is still not a practical approach for detecting many elements. For example, quadrupolar nuclei with nuclear spin > ½ possess a quadrupolar moment that causes an effective spin relaxation mechanism, causing dramatic line broadening in the NMR spectra. Unfortunately, the nuclear spin for important elements such as carbon ^{12}C isotope and ^{16}O isotope is zero; thus, no NMR signals can be detected for such nuclei. For example, the NMR active isotope of oxygen, ^{17}O has a low natural abundance (0.037%), low receptivity, quadrupolar nuclear spin = 5/2, and very short spin relaxation T1 values. These factors make ^{17}O NMR spectroscopy an unfeasible approach. Thus, nuclei with nuclear spin = ½; high natural abundance similar to those of ^{1}H, ^{31}P, and ^{19}F; and high γ are the most accessible nuclei for NMR studies in organic and biological samples. Even though the 1D ^{1}H NMR spectroscopy is the most common method, it suffers from the peak overlap as the spectra is very narrow (about 10 ppm). To overcome the challenge of spectra overlap, the multidimensional techniques which encode the indirect dimensions were proposed (i.e., Ernst, 2D, Bax 3D 90s, Clore 4D, Michal, etc.). Multidimensional NMR spectra have efficiently tackled the resolution problems for the spectra of large molecules such functional biomolecules and their respective native complexes. Applying two-dimensional (2D) or multidimensional NMR experiments are usually used to resolve overlapping peaks. Nevertheless, 2D experiments require considerably larger amounts of time than the usual 1D experiments do. The coincidence of simultaneous increase in measurement time with increase in dimensionality is directly correlated with the exact mechanism of how exactly the additional dimension is built in NMR. In an nD NMR experiment, n dimensions are created by n independent time increments which encode chemical-shift information. In 1D NMR, chemical shifts are encoded during acquisition time when data is written referred to as direct dimension hereafter. There we have no restriction in putting the number of complex points as it will not increase the measurement time significantly. The scenario changes drastically when moving to multidimensional NMR. In multidimensional NMR, first there is a prepare period and then a mixing period where the magnetization is transferred to another dimension. Here both homonuclear or heteronuclear magnetization can be generated which can be frequency labeled. This frequency labeling is the evolutionary period, and here the number of pints will determine the resolution of this dimension and the total measurement time. Each time increment is equivalent to record one additional experiment. Hence, for nD experiment, we can construct n dimensions, and in each dimension, some points have to be recorded to achieve the desired resolution in that dimension. This is the main reason the measurement time increases in a multiplicative manner with respect to the number of points in each dimension one needs to acquire to achieve the desired resolution. In the simplest multidimensional NMR, two-dimensional NMR thus uses two frequency axes. These are namely direct as explained earlier and indirect axis which is created by time increments after mixing. Now each time increment encodes one 1D spectra individually, and the second frequency is generated by the second Fourier transformation. Signal intensity is usually presented on the spectra as contours, bearing resemblance to topographic maps, or as an intensity plot. Thus, some studies (such as untargeted metabolomics) do not fully utilize 2D NMR because there are a large number of samples that would require more time to complete. Several approaches have been continuously developed to reduce the experimental time and to enhance the NMR sensitivity. These approaches include hyperpolarization methods such as dynamic nuclear polarization (DNP), cryoprobes, and new ultrahigh magnetic fields. These methods, however, are costly and need substantial hardware additions.

Other methods such as cross-polarization, INEPT, DEPT, and fast methods provide good enhancements without hardware additions and are available in most existing NMR systems. Fast 2D NMR methods and band-selective optimized-flip-angle short-transient (SOFAST) 2D NMR methods provide good NMR spectra with a short experimental time, which in turn enables the researchers to carry out powerful 2D NMR experiments in a few minutes. Increasing the number of spectral dimensions increases the spectra acquisition time dramatically. In order to overcome this issue and reduce the demand for the experimental time needed to perform the spectra measurement, different methods of acquisition of the spectra have been proposed. In this chapter, we introduce the relaxation mechanisms, and then we briefly discuss a few methods of fast methods including SOFAST, ultrafast (UF) 2D NMR, pure shift NMR, and post-processing methods. The pure shift NMR method produces spectra that only contain chemical-shift data and is void of any coupling information. This reduces the spectral complexities to a large extent as all multiplets collapse in singlets. This saves a lot of analysis time. Pure shift as such is not a fast method. However, we have incorporated this as it is indeed an important milestone in achieving high spectral resolution and a very important step towards minimizing the spectral analysis time.

2. Fast 2D NMR approaches

The power of NMR is well realized because it is the only noninvasive technique which has an atomic level resolution and can work well for both solution and solid-state samples. Although one-dimensional NMR in principle contains a wealth of information, the situation becomes severely limited due to signal overlap for small organic molecules onwards. This requires spreading the signals in other dimensions which can be the same nuclei or different NMR active nuclei. Here, concepts of 2D or multidimensional NMR arise, and with the development of 2D, 3D, and even higher-dimensional NMR, the overlap problem is solved even for large biomolecules. The price we pay is an increase in measurement time. Any higher-dimensional spectra require certain time increments to achieve certain spectral resolution, and as a result, the measurement time increases enormously as the number of dimensions increases. Generally, for the same sample moving from 1D NMR to 2D, the measurement time increases by at least two orders of magnitudes. Moving from 2D to 3D the experimental time increases further by more than one order of magnitude. Similarly, 4D and 5D become so time-consuming that they are hardly used in practice. This is the main reason for steady research in the direction to overcome the time limitation so that speedy acquisition can be feasible. Here we will discuss some of these advancements and will limit the discussion to 2D. The basic building block of the 2D or any multidimensional experiment consists of a preparation period, evolution period that incorporates incremental time delay, mixing period, and one recycle delay which is required for getting longitudinal magnetization back to equilibrium. There have been numerous methodological developments focusing on each aspect of these building blocks. Nonuniform sampling (NUS) reduces the number of incremental delays. FAST or SOFAST methods reduce recycle delay which is on the order of seconds. Ultrafast works on preparation and mixing components, and with the help of slice selection gradient, single-scan experiments are designed. We will go through each aspect separately. Lastly, there is pure shift NMR which focuses on enhancement of resolution.

2.1 SOFAST 2D NMR

We first focus on SOFAST method that aims to reduce the inter-scan delay and eventually manages to lower the 2D acquisition times [84]. This technique is optimized in order to obtain fast and sensitive 2D NMR HMQC with ^{1}H-^{15}N and ^{1}H-^{13}C correlation of biomolecules with a different size range [84]. The SOFAST-HMQC experiment makes use of selective ^{1}H pulses affecting only a portion of proton spins while keeping a large pool of proton spins unperturbed at Z axis. These unperturbed spins enhance the spin-lattice relaxation (T1) rate via dipolar interactions with the perturbed spins and effective recycle delay decreases. Additionally, selective excitation is achieved with Ernst angle excitation which helps in acquiring more sensitivity [85] at lower recycle delay. Therefore, the concerted use of Ernst angle excitation and higher repetition rates of the pulse sequence yield high signal-to-noise ratio compared to conventional experiments. This strongly is recommended in biomolecular NMR, to study protein modifications and real-time molecular kinetics [85] (**Figure 2**). Numerous versions of SOFAST-HMQC experiments have been proposed for different purposes depending on the experimental conditions [86, 87]. As an example, ^{13}C methyl SOFAST experiment has been developed, allowing the recording of high-quality methyl ^{1}H-^{13}C correlation spectra of protein with high

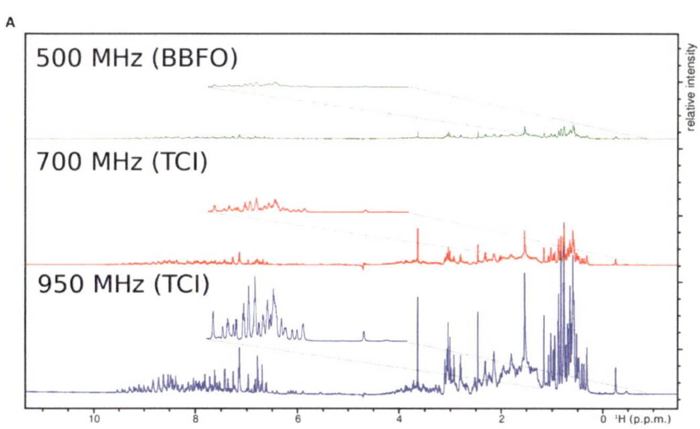

Figure 2.
Profits of the high-field NMR magnetic fields on spectral resolution and sensitivity. High-resolution 1D 1H NMR spectral of a representative small well-folded protein (A) U-15 N human ubiquitin of 1 mM concentration in 20 mM phosphate buffer, 10% D_2O, and 0.02% NaN_3 recorded with the same parameter set at three spectrometers working at 500, 700, and 950 MHz proton frequencies at 24°C.

molecular weight in a few second acquisition time [88]. Methyl groups are used as spectroscopic probes of protein structure, dynamics, and kinetics because they are dispersed naturally throughout regions of folded protein. It has been shown that the use of methyl groups as a probe is very useful, particularly for drug binding and molecular interactions studies [88].

For biomolecular NMR applications, the use of longitudinal relaxation optimized experiments permits the acceleration of 2D NMR data acquisition which provides a tool for high-throughput screening for macromolecules like proteins or nucleic acids. Such experiments allow one to obtain a high signal-to-noise ratio compared to its counterpart, as mentioned previously. Theillet et al. present an example of the use of SOFAST-HMQC experiment in order to monitor protein phosphorylation reactions [89]. In this study, the N-terminal transactivation domain of human p53 ^{15}N-labeled protein was phosphorylated by either recombinant enzyme or endogenous kinases present in cell extracts. The protein phosphorylation reactions then were monitored by SOFAST-HMQC NMR experiments in real time to obtain a high atomic-resolution understating of phosphorylations level of serine and threonine residues in kinase reactions [89].

2.2 Ultrafast 2D NMR

Secondly, we focus on ultrafast 2D NMR that is able to deliver any kind of 2D NMR spectra involving homo- or heteronuclear correlation in a single scan and, therefore, in a fraction of a second. The basis of this approach relies on the concept of revealing the spin interactions measured by spatiotemporal encoding [90]. The strategy is essentially that, instead of measuring N1 successive experiments multiple times on the sample with an independent increase of time, a better way to do it, as in the case of UF spectroscopy, is the one that "divides" the sample into N1 fractions and records all of them simultaneously within one scan (**Figure 3**) [90]. The application of slice-selective gradients encodes different phases in different slices similarly to incremental decay. As a result, the entire spectrum can be built from a single scan. However, the cost is reduced sensitivity.

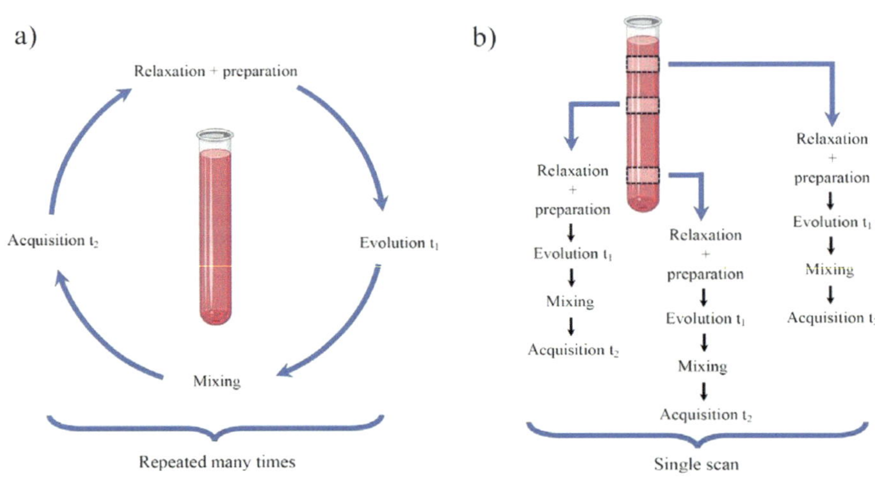

Figure 3.
Comparison between the conventional and UF 2D NMR approaches. (a) Conventional approach of 2D NMR is done on the sample by measuring N1 successive experiments multiple times with an independent increase of time. (b) UF 2D NMR approach can be conceptualized as "dividing" the sample to fractions and recording all of them simultaneously within one scan [adapted and modified from ref. [90]].

The so-called fast 2D NMR spectroscopy has found multiple applications in the field of metabolomics [86, 91, 92], as well as in the analysis of natural products [92–94]. The ultrafast quantitative 2D NMR spectroscopy is also an appropriate choice for measuring ^{13}C enrichments in a single scan (with ultrafast COSY and zTOCSY experiments). The authors of this study have reported an accuracy of 1–2%, an average precision of 3%, which is believed to be better than those previously achieved with different approaches [95]. This approach is also used in acquiring R1 longitudinal relaxation rates [4].

2.3 Nonuniform sampling and processing

Multidimensional experiments have two significant limitations: sensitivity and low resolution, especially for large molecules. For obtaining good sensitivity, we need a very high number of scans for ^{13}C, ^{15}N-labeled samples, and J-coupling transfer, and dipole-dipole interactions were utilized. For instance, metabolomics research is based on the analysis of the natural abundance of carbon and hydrogen nuclei; therefore, the sensitivity remains the main issue because isotope labeling cannot be applied. Enhancing the sensitivity and reducing the acquisition time may require the concerted use of methods such as SOFAST method and nonuniform sampling while decreasing the number of recorded points for indirect dimensions. Sometimes the single-scan approach can be used [96–98]. The main limitation of single-scan methods, however, is recording only ^{1}H plane, even for 2D experiments [99].

Some studies demonstrated that a combination of fast methods with reducing the number of data points can benefit multidimensional NMR spectra like BEST-Trosy, which provide magnification of sensitivity and resolution for structural and kinetic investigation of biomolecular and complex systems [100].

Normal 2D experiments yield 2D spectra, while during data acquisition in the indirect dimension (marked as t1), the second "vertical" dimension has incremental time delay. This time delay is linear, and the gap between the time points is the same. In contrast, NUS approaches the t1 change in a semi-random way. In NUS, only some of the time increments are recorded, and then the complete spectra are constructed after reconstructing the missing data points. Various algorithms are designed to select which data points to acquire and what will lead to successful reconstruction. The advantage of NUS is that the same level of resolution can be achieved with significantly fewer t1 increments. Thus, overall measurement time decreases. This is practical for NMR spectroscopist because this means one can obtain well-resolved 2D NMR spectra in less time.

There has been extensive research on the optimization of finding suitable combinations for sampling schemes and reconstruction algorithms [98]. NUS sampling conventionally selects random acquisition points in the indirect dimensions. The "randomness" of NUS is restricted by factors such as the transverse relaxation rate. Exponentially weighted random sampling is designed to overcome the T2 decay as more signals decay incrementally in the initial phase. Poisson gap sampling has also been used to yield excellent results as described later. Processing the recorded data can be classified in several ways [101]. For example, the sparse time domain data may be directly transformed with Fourier transform (FT) algorithms, and subsequently the missing data are added by interpolation and then followed by conventional Fourier transform [101] in the second dimension. This can also be reversed for reconstruction, followed by FT in both dimensions.

NUS can be a good tool, especially for triple resonance experiments, which have a low dynamic range. An almost identical intensity and a small number of peaks allow the use of low-density sampling. For a highly dynamic range of signals in experiments like NOESY (which provides information on the restraint through the space between atoms), reconstruction will be better with less mixing time. The positive results of reconstructed spectra were demonstrated with 20% and more sparse incremental time [102]. NUS methods have also been widely used for resonance backbone and side-chain assignments of macromolecules, determination of cross-correlated relaxation rates (CCR), kinetic processes, posttranslational modifications, and metabolomics [103].

The determinative criteria of spectra quality have a big gap in the random sampling schedule, especially at the beginning and at the end of dataset points. To overcome this limitation, the Poisson distribution [104] was used with sinusoidal variation of average gap length. In particular, sinusoidal variation of gap length keeps a degree of order in the sampling schedule, and it condenses points at the beginning of the time domain data. This provides significant benefits for time domain data with exponential decay [104].

The approach for the reconstruction of traditional uniform sampling is FT [105]. The processing spectra from NUS acquisition experiments, however, demand different types of reconstruction methods because regular discrete FT leads to artifacts as several data points are missing. Among the reconstruction algorithms, multidimensional decomposition (MDD) is the model where the full matrix of different dimensions can be identified as a sum of one-dimensional vectors (components), in which each of the components is coherent to a peak or group of peaks. MDD essentially breaks multidimensional data into sets of one-dimensional data, which are much easier to analyze and solve. The overlap in multidimensional NMR is common, and since MDD is able to resolve the overlapping resonances, MDD is well equipped for resolution enhancement of crowded datasets. MDD typically only needs a subset (20–30%) of the full dataset in order to reconstruct the full NMR signal [106, 107]. **Figure 4** gives a visual demonstration of the sampling of the full dataset to make a smaller dataset and create the spectra. One paper reports a method demonstrated on nonuniformly sampled [16] N-NOESY-HSQC datasets recorded for the 14 kDa protein azurin [108]. MDD typically only needs a subset (20–30%) of the full dataset in order to reconstruct the full NMR signal [106, 107].

Another algorithm is compressed sensing (CS) which is one of the best reconstruction methods available. Maximum entropy method (Max Ent) is another algorithm. CS is very useful in under sampled multidimensional experiments for solid-state NMR, fast metabolomics studies, and chemical-shift imaging [109, 110]. CS was demonstrated to give better reconstruction of weaker peaks, compared to an existing Max Ent implementation [110]. In one paper, two time-equivalent 3D

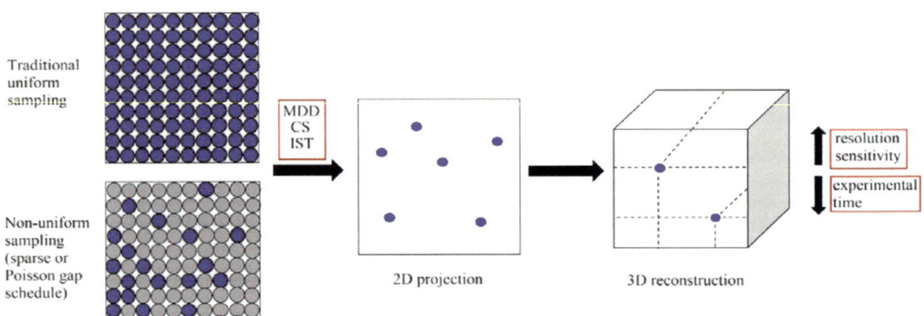

Figure 4.
Visual demonstration of NUS for obtaining NMR spectra. NUS takes a small subset of the total dataset (crowded blue dots) and uses that comparatively small dataset (20–30% of the total dataset) to create an accurate NMR spectra of the molecule of interest.

NUS TROSY-HNCA semi-constant time experiments were recorded, with CS giving better overall peak resolution [111].

Iterative soft threshold (IST) is another type of reconstruction algorithm. Like other methods, IST is able to "clean up" the NMR data by filtering the spectra. It does this by setting a threshold value and then, as its name suggests, runs through a fixed number of iterations until the procedure is complete. The best approach to IST is to use it in sync with an optimized NUS schedule. IST computational methods only take a reasonable amount of time and greatly reduce the analysis time required to obtain NMR spectra [112, 113]. One paper reports successful application of the IST approach to the ubiquitin protein. On a dual core, 2.2GHz CPU, the overall time required to process the sampled NMR dataset took under 60 seconds [114], a significant contribution to decreasing total analysis time.

3. Pure shift NMR

Proton NMR suffers from extensive signals splitting due to the presence of many weak and strong couplings, leading to significant spectral overlap, even for small molecules. The problem is severe at low magnetic fields. Although the coupling constant gives useful structural information, difficulty in obtaining them due to strong overlap often becomes a major roadblock. Ideally, a spectrum, where all multiplets collapse in singlets, gives the most straightforward information which can be achieved by pure shift. Subsequently, the analysis of couplings can be introduced in a systematic manner. The problem is simple for ^{13}C NMR as natural isotopic abundance is low and only singlets are present. However, the situation becomes complex for ^{13}C isotope-enriched samples, especially for large biomolecules. In those cases, spin-state-selective decoupling are most commonly used. Both components of the splitting due to ^{13}C-^{13}C coupling are recorded in a linear combination of inphase mode and antiphase mode. This is known as IPAP (inphase-antiphase) [115] or direct excitation of single-quantum coherences [116]. This method mainly works because of clear ^{13}C signals in different spectral shift regions and different and distinct spin-spin coupling values between those different classes of spins. This is clearly not the case for proton NMR where coupling values spread much wider ranges and also spectral overlap is more severe.

Pure shift NMR focuses on chemical shift and removes homonuclear couplings which lead to complex multiplet structure of individual peaks. This results in severe overlap even for small molecules and particularly for ^{1}H spectra since the chemical-shift dispersion of the proton is normally lower (~10 ppm) compared to other heteronuclei like ^{13}C or ^{15}N. Having the highest gyromagnetic ratio and natural abundance, ^{1}H NMR is the most obvious choice for almost all routine and conventional small and large molecule analysis. However, the spectral overlap due to homonuclear coupling limiting the resolution presents a major limitation of 1D proton NMR spectroscopy. Pure shift NMR was developed to simplify the overlapped proton NMR spectra by keeping only chemical-shift information. There have been great contributions coming first from Zangger [117–120] and subsequently by Morris and others [117, 121–127] in this direction. Some of the recent advances are presented here.

First experimental demonstration of homonuclear broadband decoupling in proton NMR was reported by Ernst and coworkers [128]. The basic pulse program consists of a 90° excitation pulse followed by spin echo. So, in direction dimension, homonuclear coupling is active, and chemical-shift evolution refocuses at the end of t1, while both chemical-shift and homonuclear couplings are active during acquisition. So, the resultant spectrum shows tilted multiplets where the indirect dimension spreads out the J-coupling. Now, 45° projection of the 2D spectrum yields pure shift spectra where all the scalar couplings have been removed. The main problem of J-resolved spectra is the twisted phase accumulation during t1 which leads to both absorptive and dispersive phase components in the indirect dimension. This results in partial or complete cancelation after taking the projection. To avoid this, normally absolute value mode calculation or power spectrum is taken before the calculation 45° projection. This generates broad peaks with long tails in the final decoupled spectrum.

The most prominent contribution came from Zangger and Sterk [120] who employed selective pulses and gradient encoding simultaneously. In this approach, first a selective 90° pulse was applied, followed by an incremental delay and combination of hard and soft selective 180° pulse with the same incremental delay and subsequent acquisition. The pulses were applied with weak gradient encoding implying spatial resolution in the sample. Hence, the whole sample is now divided in different slices, and the first frequency-selective pulse excites signals of different frequencies in different slices of the sample tube. Most importantly all signals are excited at once, but the price we pay is loss in sensitivity as the effective signal is coming from a single slice. Normally, the first experiment proposed was ~2% sensitive compared to conventional 1D NMR. In each slice, selective signals get excited by the selective first excitation pulse. Subsequently, coupling with all other spins evolves in the incremental delay. Next the selective spins undergo complete 360° rotation, while all other spins undergo 180° rotation initiated by hard 180° pulse. Therefore, after the same incremental delay, the J-coupling between the selected spin with all other spins gets refocused and homonuclear decoupling is achieved. The initially proposed scheme recorded FID with different incremental time delays. Then, from each FID, the first few points (tc/DW) determined by the incremental delay (tc) and the dwell time (DW) were concatenated to generate the final FID. The excitation bandwidth is determined by the relative proximity of the scalar-coupled protons. If coupled protons with very similar chemical shifts need to be resolved, then the selective pulses must be highly selective. This also implies the application of higher gradient strengths to generate more slices to cover the full spectral width which decreases sensitivity immensely as the effective slice thickness reduces. There have been considerable developments to overcome the limitations in the first proposed scheme.

First the two-dimensional mode of data acquisition was omitted, and the homonuclear broadband decoupling was performed during acquisition [119]. This generates a single-scan decoupling sequence like normal 1D NMR. This simplified the whole thing as these signals can be stored as regular 1D or as acquisition dimension of multidimensional spectra. Most importantly, the processing of the data becomes a routine, and there is a huge reduction of total measurement time. In this modified scheme shown in **Figure 5**, the acquisition is interrupted approximately every 1/3(^3JHH) to incorporate decoupling block. Additionally the decoupling of the signals should be placed right in the middle of the data chunk. So the first and the last acquisition block length is only half as long of all other blocks sandwiched between two.

Since relaxation is always there, some chunking artifacts arise after Fourier transform as FID gets interrupted in between. Further, to reduce the overall measurement time, this slice-selective excitation was combined with fast pulsing by shifting the frequency of the selective pulse between individual scans [117]. In slice-selective decoupling, in each slice only the excited magnetization is used for the acquisition. So, by shifting the offset of the 90° excitation pulse and also the selective 180° refocusing pulse after every scan, the unused equilibrium magnetization can be used without any recycle delay. The next scan or the subsequent scans with different offset act as recycle delay for the first set of spins which were excited in the first scan. This makes repetition much faster, and overall signal-to-noise ratio

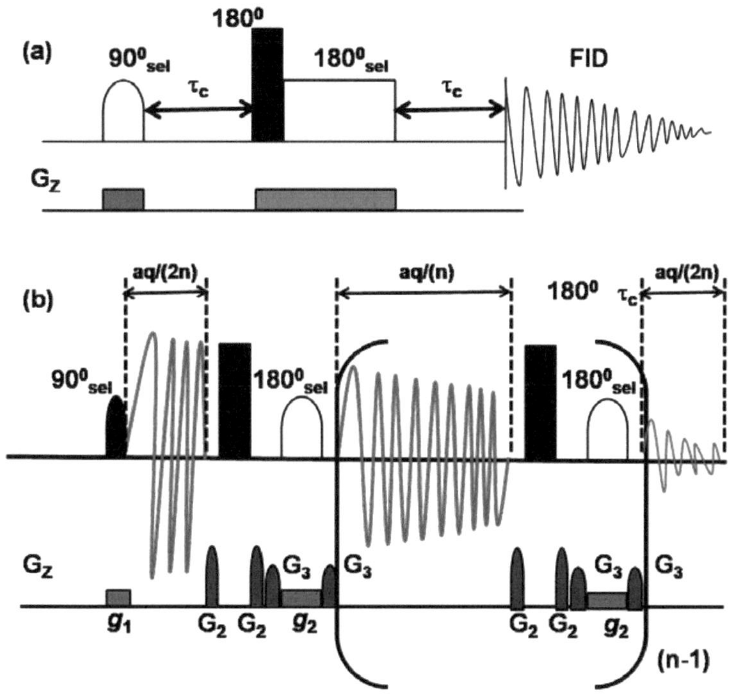

Figure 5.
(a) Pulse sequence of first Zangger-Sterk experiment. The delay τ c was incremented by several milliseconds. The first of each FID was concatenated to produce the final FID as outlined in the text. (b) the real-time slice-selective broadband homodecoupling approach. The data chunking times are variable which is represented by number n. frequency shifting is achieved for all selective pulses in parallel after each transient. The slice-selective gradients are represented by g1 and g2; G2 and G3 clear any imperfection of the refocusing pulse.

increases per unit time as shown in **Figure 6**. Apart from varying offset in between different transients, FID chunk size also varied to suppress the decoupling side-bands. The decoupling sidebands arise due to truncation of FID, and the longer the FID chunk, the more dominant are the artifacts as more antiphase magnetization can build up. On the contrary, if keeping the individual FID blocks, short leads to broad signals as more relaxation losses occur during the more frequent acquisition interruptions. The variation of chunking time between subsequent scans leads to smearing out of the decoupling sidebands which otherwise add up if they are kept the same. The slice-selective decoupling has been employed for two-dimensional homonuclear TOCSY [127, 129] and NOESY [122] spectra. The enhanced resolution in the direct dimension can be propagated to the indirect dimension by applying covariance processing which yield pure shift spectra in both the dimension for TOCSY spectrum. Additionally for heteronuclear HSQC, the slice-selective excitation was employed for intrinsically disordered proteins (IDPs) [130] where proton spectral dispersion is severely low. Additionally slice-selective excitation has been used for fast data acquisition [131] and was able to monitor chemical reaction from different spatial location of the sample tube [132].

Recently a new method named as pure shift yielded by chirp excitation (PSYCHE) experiment has been proposed by Morris et al. [125] which is based on anti Z-COSY experiment. This experiment employs two small flip angle chirp pulses which selectively refocus a small proportion of spins, namely, active spins, while leaving the majority ones undisturbed, namely, passive spins. Now this selection of active and passive spins are purely statistical which leads to an obvious advantage over ZS method as there the selection was based on selective pulse. The sequence is closely related to anti Z-COSY except to the fact that instead of using two hard small

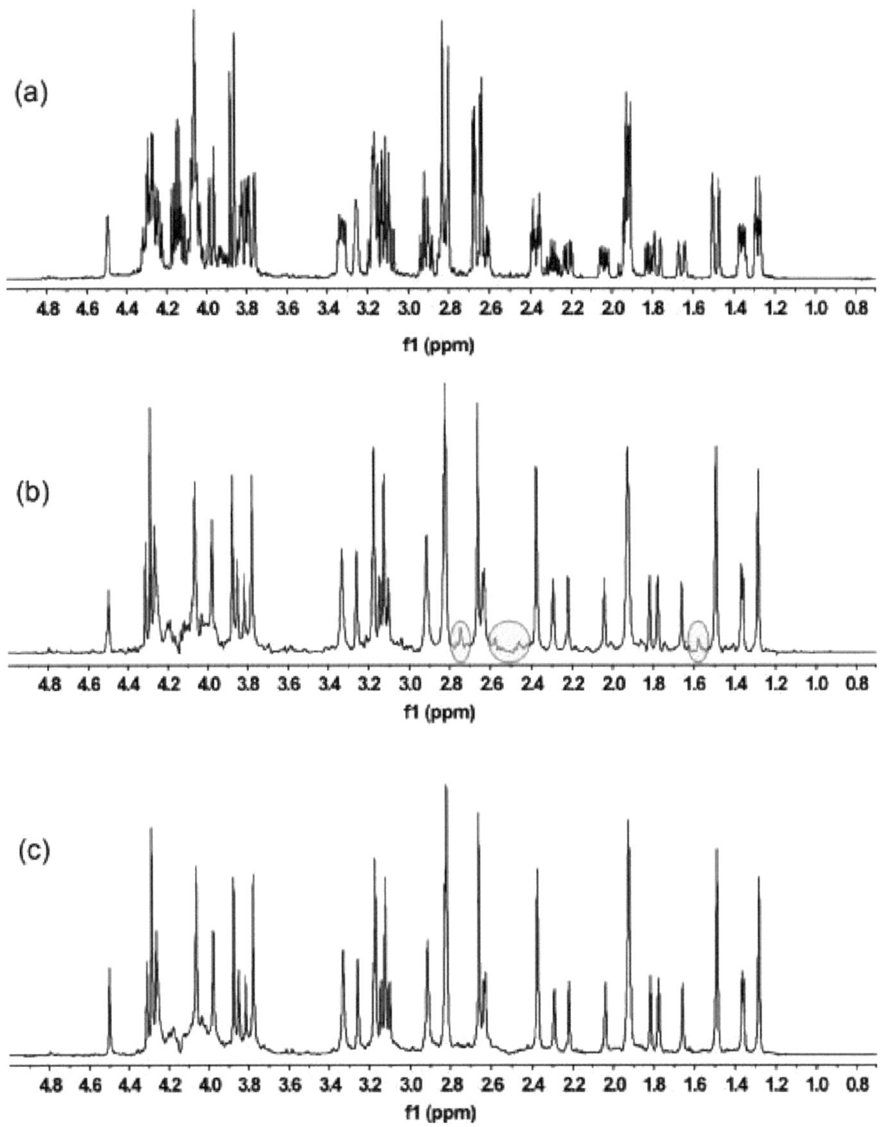

Figure 6.
(a) A regular 1D spectrum, (b) pure shift with fixed chunking time of 25 ms, (c) pure shift with variable chunking time with real-time acquired spectrum of a mixture of strychnine with unknown degradation products in $CDCl_3$. Gray circles indicate artifacts resulting from decoupling sidebands. Weak peaks visible in (b) result from additional minor compounds in the mixture, which are way below the level of artifacts in the fixed chunking time spectrum. The figure presented is from [117].

flip angle pulses, it uses two symmetric, low power, frequency-swept chirp pulses along with weak pulsed-field gradient. This gradient employs spatial resolution and suppresses the cross-peak terms as related to anti Z-COSY since they experience different chemical shifts as they are excited at different times during the chirp pulses. The major advantage of PSYCHE is that it is most sensitive among the existing pure shift methods. Several further developments are achieved to obtain ultraclean artifact-free spectra and applied in 2D NMR as well [133, 134]. Very recently one such modification named SAPPHIRE-PSYCHE methodology is applied for plant metabolomics study. In that report, single-pulse, PSYCHE, and SAPPHIRE-PSYCHE spectra were compared from aqueous extracts of *Physalis peruviana* fruits. It was found using pure shift methodology; many proton NMR signals can be cleanly observed which were not possible in conventional NMR. These signals belong to amino acids, organic acids, and sugars which are critical components of plant mixture. One example of such metabolite was the identification of glutamic acid from Cape gooseberry which was not possible to isolate due to heavy overlap. Thus, ultra-clean pure shift spectra look very promising and provide new benchmark for metabolomics studies.

4. Summary and future perspectives

High-resolution NMR with high sensitivity is the desired aim for any NMR application. There have been significant hardwire advances in achieving high sensitivity. This includes the development of high magnetic field, cryogenically cooled probes [135], recent advent of high-mass-sensitive probe for small volume [136], etc. A combination of these advances results in an increase of sensitivity by more than an order of magnitude. These advancements fueled the developments of fast methodology. Additionally due to the sample instability, which results in the relatively fast decomposition of the biomolecule being an object of the study, the quick methods of 2D spectra acquisition needed to be developed. In this regard, previously mentioned SOFAST, NUS, etc. methodologies were developed and subsequently applied in many different scenarios [90, 93, 94, 137]. Above that, these fast methods can be also used for studying either chemical or biochemical processes that are characterized by the relatively fast kinetics and also carry some structural information, like hydrogen/deuterium (H/D) exchange [138, 139]. With NUS approach, one problem is the generation of very large file size, especially for 4D and 5D experiments. Artifacts are also produced, which generates "unwanted" signals that do not accurately represent the studied molecule. However, there is considerable effort in the scientific community to reduce the amount of produced artifacts. For example, newly proposed MUNIN uses three-way decomposition and a simplified model for NMR spectra based on generally accepted assumptions. This method achieves high-resolution and good sensitivity while avoiding artifacts [106]. Also, a signal separation algorithm is shown to suppress sampling artifacts in high-resolution four-dimensional NMR spectra [140]. It is likely that NUS-based methods will continue to be used in the future, mainly because they will greatly reduce experimental time, and still provide satisfactory results with increased resolution of the spectra what facilitates their analysis. However, with the increase in magnetic field, resolution has been improved only two times and not more than that. Even with the availability of very high-field magnet such as 1.1 GHz and upwards which is limited, resolution was still a problem. Hence, the development of pure shift methods became necessary, and with the techniques great resolution can be achieved uniformly for all scenarios. In this regard, both the methodologies and developments in the direction of reducing measurement time and increasing resolution remain of utmost importance and have become an area of research itself.

Acknowledgements

We would like to thank King Abdullah University of Science and Technology for financial support.

Conflict of interest

The authors declare no conflict of interest.

Author details

Abdul-Hamid Emwas[1], Mawadda Alghrably[2], Samah Al-Harthi[2], Benjamin Gabriel Poulson[2], Kacper Szczepski[2], Kousik Chandra[2] and Mariusz Jaremko[2]*

1 Core Labs, King Abdullah University of Science and Technology (KAUST), Thuwal, Saudi Arabia

2 Biological and Environmental Science and Engineering (BESE), King Abdullah University of Science and Technology (KAUST), Thuwal, Saudi Arabia

*Address all correspondence to: mariusz.jaremko@kaust.edu.sa

References

[1] Emwas AH, Roy R, McKay RT, Tenori L, Saccenti E, Gowda GAN, et al. NMR spectroscopy for metabolomics research. Metabolites. 2019;**9**(7)

[2] Jackowski K. Multinuclear NMR spectroscopy in the gas phase. Journal of Molecular Structure. 2006;**786**(2-3):215-219

[3] Jameson CJ. Gas-phase NMR spectroscopy. Chemical Reviews. 1991;**91**(7):1375-1395

[4] Hajjar D, Kremb S, Sioud S, Emwas AH, Voolstra CR, Ravasi T. Anti-cancer agents in Saudi Arabian herbals revealed by automated high-content imaging. PLoS One. 2017;**12**(6):e0177316

[5] Harvey DJ, Royle L, Radcliffe CM, Rudd PM, Dwek RA. Structural and quantitative analysis of N-linked glycans by matrix-assisted laser desorption ionization and negative ion nanospray mass spectrometry. Analytical Biochemistry. 2008;**376**(1):44-60

[6] Guleria A, Misra DP, Rawat A, Dubey D, Khetrapal CL, Bacon P, et al. NMR-based serum metabolomics discriminates Takayasu arteritis from healthy individuals: A proof-of-principle study. Journal of Proteome Research. 2015;**14**(8):3372-3381

[7] Blindauer CA, Emwas AH, Holy A, Dvorakova H, Sletten E, Sigel H. Complex formation of the antiviral 9-2-(phosphonomethoxy) ethyl adenine (PMEA) and of its N1, N3, and N7 deaza derivatives with copper(II) in aqueous solution. Chemistry—a European Journal. 1997;**3**(9):1526-1536

[8] Cao M, Zhao L, Chen H, Xue W, Lin D. NMR-based metabolomic analysis of human bladder cancer. Analytical Sciences. 2012;**28**(5): 451-456

[9] Ali M, Shaw DR, Zhang L, Haroon MF, Narita Y, Emwas AH, et al. Aggregation ability of three phylogenetically distant anammox bacterial species. Water Research. 2018;**143**:10-18

[10] Asghar S, Shahzadi T, Alazmi M, Gao X, Emwas AH, Saleem RSZ, et al. Iridium-catalyzed regioselective borylation of substituted biaryls. Synthesis-Stuttgart. 2018;**50**(11):2211-2220

[11] Bin Farrukh SU, Javed I, Ather AQ, Emwas AH, Alazmi M, Gao X, et al. Synthesis and identification of novel pyridazinylpyrazolone based diazo compounds as inhibitors of human islet amyloid polypeptide aggregation. Bioorganic Chemistry. 2019;**84**:339-346

[12] Cui GX, Liew YJ, Li Y, Kharbatia N, Zahran NI, Emwas AH, et al. Host-dependent nitrogen recycling as a mechanism of symbiont control in Aiptasia. PLoS Genetics. 2019;**15**(6):e1008189

[13] Gallo A, Farinha ASF, Dinis M, Emwas AH, Santana A, Nielsen RJ, et al. The chemical reactions in electrosprays of water do not always correspond to those at the pristine air-water interface. Chemical Science. 2019;**10**(9):2566-2577

[14] Jameel AGA, Naser N, Emwas AH, Sarathy SM. Surrogate formulation for diesel and jet fuels using the minimalist functional group (MFG) approach. Proceedings of the Combustion Institute. 2019;**37**(4):4663-4671

[15] Jameel AGA, Naser N, Issayev G, Touitou J, Ghosh MK, Emwas AH, et al. A minimalist functional group (MFG) approach for surrogate fuel formulation. Combustion and Flame. 2018;**192**:250-271

[16] Jameel AGA, Van Oudenhoven V, Emwas AH, Sarathy SM. Predicting octane number using nuclear magnetic resonance spectroscopy and artificial neural networks. Energy and Fuels. 2018;**32**(5):6309-6329

[17] Aburabie J, Emwas AH, Peinemann KV. Silane-crosslinked asymmetric polythiosemicarbazide membranes for organic solvent nanofiltration. Macromolecular Materials and Engineering. 2019;**304**(1):1800551

[18] Alahmari F, Davaasuren B, Emwas AH, Costa P, Rothenberger A. Tris(ethylenediamine)nickel(II) thio-hydroxogermanate monohydrate: Synthesis, crystal structure, ^{1}H-NMR, EPR, optical and magnetic properties. Inorganica Chimica Acta. 2019;**488**:145-151

[19] Alahmari F, Davaasuren B, Emwas AH, Rothenberger A. Thioaluminogermanate $M(AlS_2)(GeS_2)_4$ (M = Na, Ag, Cu): Synthesis, crystal structures, characterization, ion-exchange and solid-state Al-27 and Na-23 NMR spectroscopy. Inorganic Chemistry. 2018;**57**(7):3713-3719

[20] Alahmari F, Dey S, Emwas AH, Davaasuren B, Rothenberger A. Layered copper thioaluminate $K_2Cu_3AlS_4$: Synthesis, crystal structure, characterization and solid-state Al-27 and K-39 NMR studies. Journal of Alloys and Compounds. 2019;**776**:1041-1047

[21] Aljuhani MA, Zhang ZY, Barman S, El Eter M, Failvene L, Ould-Chikh S, et al. Mechanistic study of Hydroamination of alkyne through tantalum-based silica-supported surface species. ACS Catalysis. 2019;**9**(9):8719-8725

[22] Davaasuren B, Emwas AH, Rothenberger A. MAu_2GeS_4-chalcogel (M = Co, Ni): Heterogeneous intra- and intermolecular hydroamination catalysts. Inorganic Chemistry. 2017;**56**(16):9609-9616

[23] Mallick A, El-Zohry AM, Shekhah O, Yin J, Jia JT, Aggarwal H, et al. Unprecedented ultralow detection limit of amines using a thiadiazole-functionalized Zr(IV)-based metal-organic framework. Journal of the American Chemical Society. 2019;**141**(18):7245-7249

[24] Simoes FRF, Batra NM, Emwas AH, Costa P. Validation of alkaline oxidation as a pre-treatment method for elemental quantification in single-walled carbon nanotubes. Analytical Methods. 2019;**11**(14):1884-1890

[25] Brauckmann JO, Zolfaghari P, Verhoef R, Klop EA, de Wijs GA, Kentgens APM. Structural studies of polyaramid fibers: Solid-state NMR and first-principles modeling. Macromolecules. 2016;**49**(15):5548-5560

[26] Cabrera Y, Cabrera A, Larsen FH, Felby C. Solid-state Si-29 NMR and FTIR analyses of lignin-silica coprecipitates. Holzforschung. 2016;**70**(8):709-718

[27] Elkins MR, Wang T, Nick M, Jo H, Lemmin T, Prusiner SB, et al. Structural polymorphism of Alzheimer's beta-amyloid fibrils as controlled by an E22 switch: A solid-state NMR study. Journal of the American Chemical Society. 2016;**138**(31):9840-9852

[28] Lang S, Benz M, Obenaus U, Himmelmann R, Hunger M. Novel approach for the characterization of Lewis acidic solid catalysts by solid-state NMR spectroscopy. ChemCatChem. 2016;**8**(12):2031-2036

[29] Chu SD, Maltsev S, Emwas AH, Lorigan GA. Solid-state NMR paramagnetic relaxation enhancement immersion depth studies in phospholipid bilayers. Journal of Magnetic Resonance. 2010;**207**(1):89-94

[30] Emwas AHM, Al-Talla ZA, Guo XR, Al-Ghamdi S, Al-Masri HT. Utilizing

NMR and EPR spectroscopy to probe the role of copper in prion diseases. Magnetic Resonance in Chemistry. 2013;**51**(5):255-268

[31] Mayo D, Zhou A, Sahu I, McCarrick R, Walton P, Ring A, et al. Probing the structure of membrane proteins with electron spin echo envelope modulation spectroscopy. Protein Science. 2011;**20**(7):1100-1104

[32] Akbey U, Oschkinat H. Structural biology applications of solid state MAS DNP NMR. Journal of Magnetic Resonance. 2016;**269**:213-224

[33] Caro JA, Wand AJ. Practical aspects of high-pressure NMR spectroscopy and its applications in protein biophysics and structural biology. Methods. 2018;**148**:67-80

[34] Ciambellotti S, Turano P. Structural biology of iron-binding proteins by NMR spectroscopy. European Journal of Inorganic Chemistry. 2019;**5**:569-576

[35] Dass R, Grudziaz K, Ishikawa T, Nowakowski M, Debowska R, Kazimierczuk K. Fast 2D NMR spectroscopy for in vivo monitoring of bacterial metabolism in complex mixtures. Frontiers in Microbiology. 2017;**8**:1306

[36] Malar AA, Smith-Penzel S, Camenisch GM, Wiegand T, Samoson A, Bockmann A, et al. Quantifying proton NMR coherent linewidth in proteins under fast MAS conditions: A second moment approach. Physical Chemistry Chemical Physics. 2019;**21**(35):18850-18865

[37] Opella SJ, Marassi FM. Applications of NMR to membrane proteins. Archives of Biochemistry and Biophysics. 2017;**628**:92-101

[38] Rahman S, Byun Y, Hassan MI, Kim J, Kumar V. Towards understanding cellular structure biology: In-cell NMR. Biochimica et Biophysica Acta - Proteins and Proteomics. 2017;**1865**(5):547-557

[39] Al-Masri HT, Emwas AHM, Al-Talla ZA, Alkordi MH. Synthesis and characterization of new N-(diphenylphosphino)-naphthylamine chalcogenides: X-ray structures of $(1\text{-}NHC_{10}H_7)P(Se)Ph\text{-}2$ and $Ph_2P(S)OP(S)Ph\text{-}2$. Phosphorus Sulfur Silicon and the Related Elements. 2012;**187**(9):1082-1090

[40] Atiqullah M, Al-Harthi MA, Anantawaraskul S, Emwas AHM. Ethylene homo- and copolymerization chain-transfers: A perspective from supported (nBuCp)(2)$ZrCl_2$ catalyst active Centre distribution. Journal of Chemical Sciences. 2015;**127**(4):717-728

[41] Batool F, Parveen S, Emwas AH, Sioud S, Gao X, Munawar MA, et al. Synthesis of fluoroalkoxy substituted arylboronic esters by iridium-catalyzed aromatic C-H borylation. Organic Letters. 2015;**17**(17):4256-4259

[42] Jameel AGA, Elbaz AM, Emwas AH, Roberts WL, Sarathy SM. Calculation of average molecular parameters, functional groups, and a surrogate molecule for heavy fuel oils using H-1 and C-13 nuclear magnetic resonance spectroscopy. Energy and Fuels. 2016;**30**(5):3894-3905

[43] Jameel AGA, Naser N, Emwas AH, Dooley S, Sarathy SM. Predicting fuel ignition quality using H-1 NMR spectroscopy and multiple linear regression. Energy and Fuels. 2016;**30**(11):9819-9835

[44] Li S, Winters H, Jeong S, Emwas AH, Vigneswaran S, Amy GL. Marine bacterial transparent exopolymer particles (TEP) and TEP precursors: Characterization and RO fouling potential. Desalination. 2016;**379**:68-74

[45] Blumich B, Singh K. Desktop NMR and its applications from materials science to organic chemistry. Angewandte Chemie International Edition. 2018;**57**(24):6996-7010

[46] He TY, Musah RA. Multidimensional high-resolution NMR structural characterization of a carborane cluster derivative: The case of 2-amino-3-(1,7-dicarba-closo-dodecaboranyl-1-thio)propanoic acid. Polyhedron. 2019;**163**:171-177

[47] Saito T, Yamazaki T, Numata M. Development of nuclear magnetic resonance as a tool of quantitative analysis for organic materials. Metrologia. 2019;**56**(5):054002

[48] Yu Y, Rebek J. Reactions of folded molecules in water. Accounts of Chemical Research. 2018;**51**(12):3031-3040

[49] Campbell K, Ooms KJ, Ferguson MJ, Stang PJ, Wasylishen RE, Tykwinski RR. Shape-persistent macrocycles - self-assembly reactions and characterization by hyperpolarized Xe-129 NMR spectroscopy. Canadian Journal of Chemistry. 2011;**89**(10):1264-1276

[50] Figueiredo S, Gomes AC, Neves P, Amarante TR, Paz FAA, Soares R, et al. Synthesis, structural elucidation, and application of a pyrazolylpyridine-molybdenum oxide composite as a heterogeneous catalyst for olefin epoxidation. Inorganic Chemistry. 2012;**51**(15):8629-8635

[51] Primikyri A, Kyriakou E, Charisiadis P, Tsiafoulis C, Stamatis H, Tzakos AG, et al. Fine-tuning of the diffusion dimension of -OH groups for high resolution DOSY NMR applications in crude enzymatic transformations and mixtures of organic compounds. Tetrahedron. 2012;**68**(34):6887-6891

[52] Rudzinska E, Dziedziola G, Berlicki L, Kafarski P. Enantiodifferentiation of alpha-hydroxyalkanephosphonic acids in P-31 NMR with application of alpha-cyclodextrin as chiral discriminating agent. Chirality. 2010;**22**(1):63-68

[53] Atiqullah M, Adamu S, Emwas AHM. UHMW Ziegler-Natta polyethylene: Synthesis, crystallization, and melt behavior. Journal of the Taiwan Institute of Chemical Engineers. 2017;**76**:141-155

[54] Adamu S, Atiqullah M, Malaibari ZO, Al-Harthi MA, Emwas AHM, Ul-Hamid A. Metallocene-catalyzed ethylene-alpha-olefin isomeric copolymerization: A perspective from hydrodynamic boundary layer mass transfer and design of MAO anion. Journal of the Taiwan Institute of Chemical Engineers. 2016;**60**:92-105

[55] Alkordi MH, Haikal RR, Hassan YS, Emwas AH, Belmabkhout Y. Poly-functional porous-organic polymers to access functionality - CO_2 sorption energetic relationships. Journal of Materials Chemistry A. 2015;**3**(45):22584-22590

[56] Atiqullah M, Anantawaraskul S, Emwas AHM, Al-Harthi MA, Hussain I, Ul-Hamid A, et al. Effects of supported ((BuCp)-Bu-n)(2)ZrCl2 catalyst active-center distribution on ethylene-1-hexene copolymer backbone heterogeneity and thermal behaviors. Industrial and Engineering Chemistry Research. 2013;**52**(27):9359-9373

[57] Atiqullah M, Anantawaraskul S, Emwas AHM, Al-Harthi MA, Hussain I, Ul-Hamid A, et al. Silica-supported ((BuCp)-Bu-n)(2)$ZrCl_2$: Effect of catalyst active center distribution on ethylene-1-hexene copolymerization. Polymer International. 2014;**63**(6):955-972

[58] Atiqullah M, Winston MS, Bercaw JE, Hussain I, Fazal A, Al-Harthi MA, et al. Effects of a

vanadium post-metallocene catalyst-induced polymer backbone inhomogeneity on UV oxidative degradation of the resulting polyethylene film. Polymer Degradation and Stability. 2012;**97**(7):1164-1177

[59] Bahuleyan BK, De SK, Sarath PU, Furquan SA, Masihullah JK, Emwas AH, et al. Effect of aluminium nitride on the properties of polyethylene obtained by In situ polymerization using Ni(II) diimine complex. Macromolecular Research. 2012;**20**(7):772-775

[60] Kamal MS, Bahuleyan BK, Sohail OB, Emwas AHM, Bercaw JE, Al-Harthi MA. Crystallization analysis fractionation of poly(ethylene-co-styrene) produced by metallocene catalysts. Polymer Bulletin. 2013;**70**(9):2645-2656

[61] Arachchilage I, Patel MK, Harmon JP. Synthesizing radiation-hard polymer and copolymers using laccol monomers extracted from lacquer tree toxicodendron succedanea via cationic polymerization. Polymer Engineering and Science. 2019;**59**(8):1611-1623

[62] Arrabal-Campos FM, Aguilera-Saez LM, Fernandez I. Algebraic reconstruction technique for diffusion NMR experiments. Application to the molecular weight prediction of polymers. Journal of Physical Chemistry A. 2019;**123**(4):943-950

[63] Brown SP. Advanced solid-state NMR methods for characterising structure and self-assembly in supramolecular chemistry, polymers and hydrogels. Current Opinion in Colloid and Interface Science. 2018;**33**:86-98

[64] Zare Y, Dabbaghi A, Rahmani S. Increasing the hydrophilicity of star-shaped amphiphilic co-networks by using of PEG and dendritic s-PCL cross-linkers. Polymers for Advanced Technologies. 2019

[65] Bosnjakovic-Pavlovic N, Bajuk-Bogdanovic D, Zakrzewska J, Yan ZY, Holclajtner-Antunovic I, Gillet JM, et al. Reactivity of 12-tungstophosphoric acid and its inhibitor potency toward Na+/K+-ATPase: A combined P-31 NMR study, ab initio calculations and crystallographic analysis. Journal of Inorganic Biochemistry. 2017;**176**:90-99

[66] Brauniger T, Jansen M. Solid-state NMR spectroscopy of quadrupolar nuclei in inorganic chemistry. Zeitschrift für Anorganische und Allgemeine Chemie. 2013;**639**(6):857-879

[67] Ha M, Thiessen AN, Sergeyev IV, Veinot JGC, Michaelis VK. Endogenous dynamic nuclear polarization NMR of hydride-terminated silicon nanoparticles. Solid State Nuclear Magnetic Resonance. 2019;**100**:77-84

[68] Marbella LE, Millstone JE. NMR techniques for noble metal nanoparticles. Chemistry of Materials. 2015;**27**(8):2721-2739

[69] Martin RW, Kelly JE, Collier KA. Spatial reorientation experiments for NMR of solids and partially oriented liquids. Progress in Nuclear Magnetic Resonance Spectroscopy. 2015;**90-91**:92-122

[70] Alezi D, Belmabkhout Y, Suyetin M, Bhatt PM, Weselinski LJ, Solovyeva V, et al. MOF crystal chemistry paving the way to gas storage needs: Aluminum-based soc-MOF for CH_4, O_2, and CO_2 storage. Journal of the American Chemical Society. 2015;**137**(41):13308-13318

[71] Dey R, Samantaray MK, Poater A, Hamieh A, Kavitake S, Abou-Hamad E, et al. Synthesis and characterization of a homogeneous and silica supported homoleptic cationic tungsten(VI) methyl complex: Application in olefin metathesis. Chemical Communications. 2016;**52**(75):11270-11273

[72] Geng GQ, Taylor R, Bae S, Hernandez-Cruz D, Kilcoyne DA, Emwas AH, et al. Atomic and nano-scale characterization of a 50-year-old hydrated C3S paste. Cement and Concrete Research. 2015;77:36-46

[73] Kirchheim AP, Dal Molin DC, Fischer P, Emwas AH, Provis JL, Monteiro PJM. Real-time high-resolution X-ray imaging and nuclear magnetic resonance study of the hydration of pure and Na-doped C(3) A in the presence of sulfates. Inorganic Chemistry. 2011;**50**(4):1203-1212

[74] Mroue KH, Emwas AHM, Power WP. Solid-state Al-27 nuclear magnetic resonance investigation of three aluminum-centered dyes. Canadian Journal of Chemistry. 2010;**88**(2):111-123

[75] Saliba D, Ezzeddine A, Emwas AH, Khashab NM, Al-Ghoul M. Dynamics and mechanism of intercalation/de-intercalation of Rhodamine B during the polymorphic transformation of the CdAl layered double hydroxide to the brucite-like cadmium hydroxide. Crystal Growth and Design. 2016;**16**(8):4327-4335

[76] Grigor'ev GY, Nabiev SS. Production and applications of spin-polarized isotopes of noble gases. Russian Journal of Physical Chemistry B. 2018;**12**(3):363-377

[77] Witte C, Kunth M, Dopfert J, Rossella F, Schroder L. Hyperpolarized xenon for NMR and MRI applications. Journal of Visualized Experiments. 2012;**67**:e4268

[78] Auccaise R, Araujo-Ferreira AG, Sarthour RS, Oliveira IS, Bonagamba TJ, Roditi I. Spin squeezing in a quadrupolar nuclei NMR system. Physical Review Letters. 2015;**114**(4)

[79] Glenn DR, Bucher DB, Lee J, Lukin MD, Park H, Walsworth RL. High-resolution magnetic resonance spectroscopy using a solid-state spin sensor. Nature. 2018;**555**(7696):351

[80] Mananga ES, Charpentier T. Revisiting the applications of Floquet-Magnus and Fer expansion approaches in physics and solid-state NMR. International Journal of Modern Physics B. 2018;**32**(22):1850236

[81] Xin T, Wang BX, Li KR, Kong XY, Wei SJ, Wang T, et al. Nuclear magnetic resonance for quantum computing: Techniques and recent achievements. Chinese Physics B. 2018;**27**(2)

[82] Emwas AH, Saunders M, Ludwig C, Gunther UL. Determinants for optimal enhancement in ex situ DNP experiments. Applied Magnetic Resonance. 2008;**34**(3-4):483-494

[83] Mattar SM, Emwas AH, Calhoun LA. Spectroscopic studies of the intermediates in the conversion of 1,4,11,12-tetrahydro-9,10-anthraquinone to 9,10-anthraquinone by reaction with oxygen under basic conditions. Journal of Physical Chemistry A. 2004;**108**(52):11545-11553

[84] Schanda P, Kupče Ē, Brutscher B. SOFAST-HMQC experiments for recording two-dimensional deteronuclear correlation spectra of proteins within a few seconds. Journal of Biomolecular NMR. 2005;**33**(4):199-211

[85] Brutscher B. SOFAST HMQC. In: GCK R, editor. Encyclopedia of Biophysics. Berlin, Heidelberg: Springer Berlin Heidelberg; 2013. p. 2407

[86] Ghosh S, Sengupta A, Chandra K. SOFAST-HMQC—An efficient tool for metabolomics. Analytical and Bioanalytical Chemistry. 2017;**409**(29):6731-6738

[87] Theillet F-X, Rose HM, Liokatis S, Binolfi A, Thongwichian R, Stuiver M,

et al. Corrigendum: Site-specific NMR mapping and time-resolved monitoring of serine and threonine phosphorylation in reconstituted kinase reactions and mammalian cell extracts. Nature Protocols. 2016;**11**(1):192

[88] Schanda P. Development and Application of Fast NMR Methods for the Study of Protein Structure and Dynamics. Biological Physics. Université Joseph-Fourier - Grenoble I. 2007. English

[89] Zhan H, Lin X, Wei Z, Ye Q, Cai S, You X, et al. A single-scan inhomogeneity-tolerant NMR method for high-resolution two-dimensional J-resolved spectroscopy. IEEE Transactions on Biomedical Engineering. 2018;**66**(6):1559-1566

[90] Giraudeau P, Frydman L. Ultrafast 2D NMR: An emerging tool in analytical spectroscopy. Annual Review of Analytical Chemistry. 2014;**7**:129-161

[91] Emwas A-H, Roy R, McKay RT, Tenori L, Saccenti E, Gowda G, et al. NMR spectroscopy for metabolomics research. Metabolites. 2019;**9**(7):123

[92] Ghosh S, Sengupta A, Chandra K. Quantitative metabolic profiling of NMR spectral signatures of branched chain amino acids in blood serum. Amino Acids. 2015;**47**(10):2229-2236

[93] Giraudeau P, Shrot Y, Frydman L. Multiple ultrafast, broadband 2D NMR spectra of hyperpolarized natural products. Journal of the American Chemical Society. 2009;**131**(39):13902-13903

[94] Guennec AL, Giraudeau P, Caldarelli S. Evaluation of fast 2D NMR for metabolomics. Analytical Chemistry. 2014;**86**(12):5946-5954

[95] Giraudeau P, Massou S, Robin Y, Cahoreau E, Portais J-C, Akoka S. Ultrafast quantitative 2D NMR: An efficient tool for the measurement of specific isotopic enrichments in complex biological mixtures. Analytical Chemistry. 2011;**83**(8):3112-3119

[96] Frydman L, Scherf T, Lupulescu A. The acquisition of multidimensional NMR spectra within a single scan. Proceedings of the National Academy of Sciences. 2002;**99**(25):15858-15862

[97] Mishkovsky M, Frydman L. Progress in hyperpolarized ultrafast 2D NMR spectroscopy. ChemPhysChem. 2008;**9**(16):2340-2348

[98] Barna J, Laue E, Mayger MR, Skilling J, Worrall S. Exponential sampling, an alternative method for sampling in two-dimensional NMR experiments. Journal of Magnetic Resonance. 1987;**73**(1):69-77

[99] Tal A, Frydman L. Single-scan multidimensional magnetic resonance. Progress in Nuclear Magnetic Resonance Spectroscopy. 2010;**57**(3):241-292

[100] Solyom Z, Schwarten M, Geist L, Konrat R, Willbold D, Brutscher B. BEST-TROSY experiments for time-efficient sequential resonance assignment of large disordered proteins. Journal of Biomolecular NMR. 2013;**55**(4):311-321

[101] Billeter M. Non-uniform Sampling in Biomolecular NMR. Journal of Biomolecular NMR. Springer. 1 June 2017;**68**(2):65-66

[102] Rovnyak D, Hoch J, Stern A, Wagner G. Resolution and sensitivity of high field nuclear magnetic resonance spectroscopy. Journal of Biomolecular NMR. 2004;**30**(1):1-10

[103] Zeng J, Zhou P, Donald BR. Protein side-chain resonance assignment and NOE assignment using RDC-defined backbones without TOCSY data. Journal of Biomolecular NMR. 2011;**50**(4):371

[104] Hyberts SG, Takeuchi K, Wagner G. Poisson-gap sampling and forward maximum entropy reconstruction for enhancing the resolution and sensitivity of protein NMR data. Journal of the American Chemical Society. 2010;**132**(7):2145-2147

[105] Ernst RR, Anderson WA. Application of Fourier transform spectroscopy to magnetic resonance. The Review of Scientific Instruments. 1966;**37**(1):93-102

[106] Orekhov VY, Ibraghimov IV, Billeter M. MUNIN: A new approach to multi-dimensional NMR spectra interpretation. Journal of Biomolecular NMR. 2001;**20**(1):49-60

[107] Tugarinov V, Kay LE, Ibraghimov I, Orekhov VY. High-resolution four-dimensional 1H– 13C NOE spectroscopy using methyl-TROSY, sparse data acquisition, and multidimensional decomposition. Journal of the American Chemical Society. 2005;**127**(8):2767-2775

[108] Orekhov VY, Ibraghimov I, Billeter M. Optimizing resolution in multidimensional NMR by three-way decomposition. Journal of Biomolecular NMR. 2003;**27**(2):165-173

[109] Kazimierczuk K, Orekhov VY. Accelerated NMR spectroscopy by using compressed sensing. Angewandte Chemie International Edition. 2011;**50**(24):5556-5559

[110] Holland DJ, Bostock MJ, Gladden LF, Nietlispach D. Fast multidimensional NMR spectroscopy using compressed sensing. Angewandte Chemie International Edition. 2011;**50**(29):6548-6551

[111] Bostock M, Nietlispach D. Compressed sensing: Reconstruction of non-uniformly sampled multidimensional NMR data. Concepts in Magnetic Resonance Part A. 2017;**46**(2):e21438

[112] Drori I. Fast minimization by iterative thresholding for multidimensional NMR spectroscopy. EURASIP Journal on Advances in Signal Processing. 2007;**2007**(1):020248

[113] Hyberts SG, Milbradt AG, Wagner AB, Arthanari H, Wagner G. Application of iterative soft thresholding for fast reconstruction of NMR data non-uniformly sampled with multidimensional Poisson gap scheduling. Journal of Biomolecular NMR. 2012;**52**(4):315-327

[114] Shrot Y, Frydman L. Compressed sensing and the reconstruction of ultrafast 2D NMR data: Principles and biomolecular applications. Journal of Magnetic Resonance. 2011;**209**(2):352-358

[115] Duma L, Hediger S, Brutscher B, Böckmann A, Emsley L. Resolution enhancement in multidimensional solid-state NMR spectroscopy of proteins using spin-state selection. Journal of the American Chemical Society. 2003;**125**(39):11816-11817

[116] Meissner A, Sørensen OW. Spin-state-selective excitation. Application for E. COSY-type measurement ofJHHcoupling constants. Journal of Magnetic Resonance. 1997;**128**(1):92-97

[117] Mauhart J, Glanzer S, Sakhaii P, Bermel W, Zangger K. Faster and cleaner real-time pure shift NMR experiments. Journal of Magnetic Resonance. 2015;**259**:207-215

[118] Sakhaii P, Haase B, Bermel W, Kerssebaum R, Wagner GE, Zangger K. Broadband homodecoupled NMR spectroscopy with enhanced sensitivity. Journal of Magnetic Resonance. 2013;**233**:92-95

[119] Meyer NH, Zangger K. Simplifying proton NMR spectra by instant

homonuclear broadband decoupling. Angewandte Chemie International Edition. 2013;**52**(28):7143-7146

[120] Zangger K, Sterk H. Homonuclear broadband-decoupled NMR spectra. Journal of Magnetic Resonance. 1997;**2**(124):486-489

[121] Aguilar JA, Nilsson M, Bodenhausen G, Morris GA. Spin echo NMR spectra without J modulation. Chemical Communications. 2012;**48**(6):811-813

[122] Aguilar JA, Colbourne AA, Cassani J, Nilsson M, Morris GA. Decoupling two-dimensional NMR spectroscopy in both dimensions: Pure shift NOESY and COSY. Angewandte Chemie International Edition. 2012;**51**(26):6460-6463

[123] Aguilar JA, Faulkner S, Nilsson M, Morris GA. Pure shift 1H NMR: A resolution of the resolution problem? Angewandte Chemie International Edition. 2010;**49**(23):3901-3903

[124] Aguilar JA, Nilsson M, Morris GA. Simple proton spectra from complex spin systems: Pure shift NMR spectroscopy using BIRD. Angewandte Chemie International Edition. 2011;**50**(41):9716-9717

[125] Foroozandeh M, Adams RW, Meharry NJ, Jeannerat D, Nilsson M, Morris GA. Ultrahigh-resolution NMR spectroscopy. Angewandte Chemie International Edition. 2014;**53**(27):6990-6992

[126] Foroozandeh M, Adams RW, Nilsson M, Morris GA. Ultrahigh-resolution total correlation NMR spectroscopy. Journal of the American Chemical Society. 2014;**136**(34):11867-11869

[127] Morris GA, Aguilar JA, Evans R, Haiber S, Nilsson M. True chemical shift correlation maps: A TOCSY experiment with pure shifts in both dimensions. Journal of the American Chemical Society. 2010;**132**(37):12770-12772

[128] Aue W, Karhan J, Ernst R. Homonuclear broad band decoupling and two-dimensional J-resolved NMR spectroscopy. The Journal of Chemical Physics. 1976;**64**(10):4226-4227

[129] Koivisto JJ. Zero-quantum filtered pure shift TOCSY. Chemical Communications. 2013;**49**(1):96-98

[130] Meyer NH, Zangger K. Enhancing the resolution of multi-dimensional heteronuclear NMR spectra of intrinsically disordered proteins by homonuclear broadband decoupling. Chemical Communications. 2014;**50**(12):1488-1490

[131] Wagner GE, Sakhaii P, Bermel W, Zangger K. Monitoring fast reactions by spatially-selective and frequency-shifted continuous NMR spectroscopy: Application to rapid-injection protein unfolding. Chemical Communications. 2013;**49**(30):3155-3157

[132] Pöppler AC, Frischkorn S, Stalke D, John M. Toluene and lithium amide diffusion into polystyrene: A slice-selective NMR-spectroscopic study. ChemPhysChem. 2013;**14**(13):3103-3107

[133] Foroozandeh M, Morris GA, Nilsson M. PSYCHE pure shift NMR spectroscopy. Chemistry - A European Journal. 2018;**24**(53):13988-14000

[134] Foroozandeh M, Castañar L, Martins LG, Sinnaeve D, Poggetto GD, Tormena CF, et al. Ultrahigh-resolution diffusion-ordered spectroscopy. Angewandte Chemie. 2016;**128**(50):15808-15811

[135] Kovacs H, Moskau D, Spraul M. Cryogenically cooled probes—A leap in NMR technology. Progress in Nuclear Magnetic Resonance Spectroscopy. 2005;**46**(2):131-155

[136] Everett JR, Harris RK, Lindon JC, Wilson ID. NMR in Pharmaceutical Science. John Wiley & Sons; 2015

[137] Giraudeau P, Akoka S. A new detection scheme for ultrafast 2D J-resolved spectroscopy. Journal of Magnetic Resonance. 2007;**186**(2):352-357

[138] Schanda P, Brutscher B. Very fast two-dimensional NMR spectroscopy for real-time investigation of dynamic events in proteins on the time scale of seconds. Journal of the American Chemical Society. 2005;**127**(22):8014-8015

[139] Gal M, Kern T, Schanda P, Frydman L, Brutscher B. An improved ultrafast 2D NMR experiment: Towards atom-resolved real-time studies of protein kinetics at multi-Hz rates. Journal of Biomolecular NMR. 2009;**43**(1):1-10

[140] Stanek J, Augustyniak R, Koźmiński W. Suppression of sampling artefacts in high-resolution four-dimensional NMR spectra using signal separation algorithm. Journal of Magnetic Resonance. 2012;**214**:91-102

Chapter 6

Solubility, Discoloration, and Solid-State ^{13}C NMR Spectra of Stereoregular Poly(Vinyl Chloride) Prepared by Urea Clathrate Polymerization at Low Temperatures

Masatomo Minagawa, Jun Yatabe, Fumio Yoshii, Shin Hasegawa, Nobuhiro Sato and Tomochika Matsuyama

Abstract

Stereoregular poly(vinyl chloride) (PVC) was obtained by urea clathrate polymerization. The sample was a white crystalline powder. Its molecular structure was studied by appearance, FT-IR, WAXD, and NMR (solid) in comparison with those of ordinary free radical one. The sample was totally insoluble to polar solvent such as DMF in contrast with good solubility of free radical one. Prolonged heating at high temperatures ensured discoloration and elimination reaction permitted formation of trans-type double bond. This structural change was traced by FT-IR and solid-state NMR. Two non-compromise characteristics, stiff molecular chain and easy discoloration, in canal PVC are described.

Keywords: stereoregular PVC, analysis by WAXD/FT-IR/NMR/ESR, elimination reaction

1. Introduction

Solid-state polymerization is a unique polymer synthetic method in Polymer Science. The most famous example is probably a urea clathrate polymerization of vinyl chloride (VC) as described here. That is, VC monomers are packed regularly in one-dimensional narrow urea canal under low temperatures (canal complex or inclusion complex). When strong γ-irradiation or electron beam one will be made, polymerization takes place, and highly stereoregular poly(vinyl chloride) (PVC) is obtained.

The first research in this area was carried out by Brown and White, researchers of GE company (USA) [1, 2]. They showed that urea canal-polymerized polymers have a remarkable difference in its physical properties. Particularly, not only ordinary urea but also thiourea does the canal complex at low temperature, and their difference is an inner diameter of cavity (5A or 6A). The structures and properties

 IntechOpen

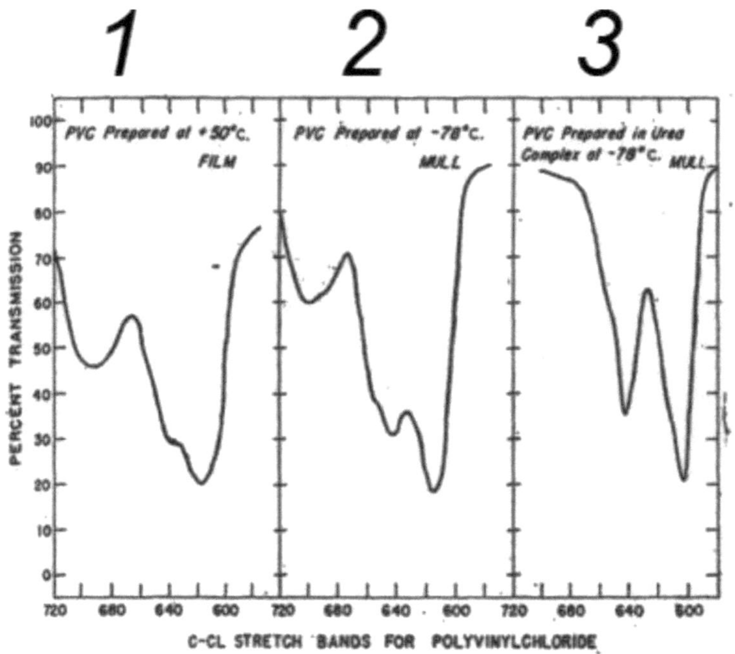

Figure 1.
Krimm's data [3, 4].

of the resulting polymers are well described by a limited number of instruments including solubility measurements, etc. Although the described time period is old, their valuable finding and observation have still a brilliant light in Polymer Science even at the present time.

Stereoregular PVC was investigated by Krimm et al. by using IR spectroscopy [3, 4]. Their first paper on stereoregular PVC was only one page, but its IR spectra were printed and published in all the textbooks and professional ones in the world (see **Figure 1**). In the IR spectra, structural difference appeared in the 700–600 cm^{-1} region. One can notice that only two peaks are clearly seen in the urea canal PVC in contrast with three peaks in free radical one. Lack of the left-hand side band (690 cm^{-1}, assigned to be isotactic) strongly suggests that stereoregular PVC is highly syndiotactic.

In Japan, a detailed IR study of PVC including stereoregular one was carried out by Shimanouchi and Tasumi in the University of Tokyo [5, 6]. Their standpoint of view was purely scientific, and direct application to industrial field appeared to be relatively small. However, their effort was greatly helpful for the improvement of commercial PVC products in Japan. They synthesized various model compounds and confirmed IR assignment of PVC, for example, the effects of C-Cl position on IR spectra and related problems are typical examples. Their collaborative work has been summarized and published in Ref. [7].

2. Theory: principle of urea clathrate polymerization

Figure 2 shows the principle of urea canal polymerization [8–10]. When organic monomers are mixed with fine urea and the mixture is kept at low temperatures (−78°C), a canal complex is formed spontaneously. This is a typical inclusion phenomenon, and the resulting canal complex is called *inclusion complex*. It must be noted that such an inclusion is caused by *the phase transition* of urea, when urea is mixed with organic molecules and then temperature is kept at such lower temperatures.

The geometrical shape and size of the complex are quite different according to the type of monomers [9, 10]. An ideal complex is n-paraffin/urea system (**Figure 2**, right). A well-known fact is that started from ethylene/urea system, linear polyethylene with no branching is obtained.

In the case of polar vinyl monomers, inclusion structure is quite different according to the type of monomers, e.g., *monoclinic* (VC/urea) and *hexagonal* (AN/urea) (**Figure 3**). That is, slurry of initial mixture (urea/liquid monomers) turned to be dry flake-like white substance eventually under such low temperatures. The γ-irradiation toward the canal complex must be carried out carefully under

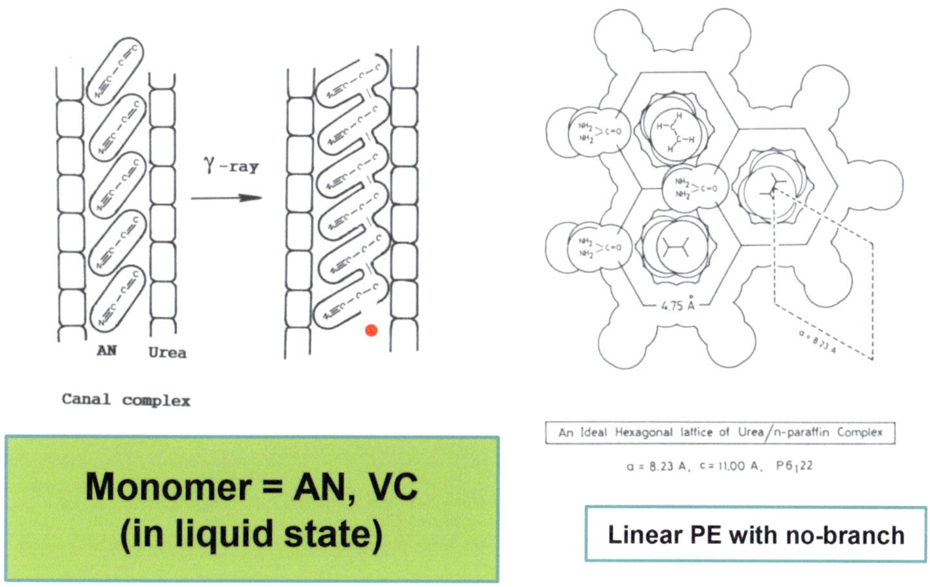

Figure 2.
Principle of urea canal polymerization.

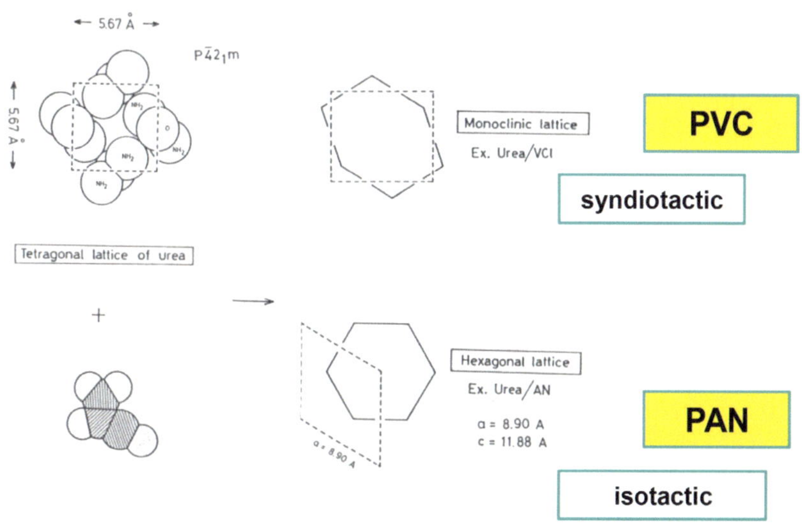

Figure 3.
The shape and size of canal complex.

sufficiently cooled conditions (below −78°C). This is because if temperature is increased, the canal complex is destroyed, then unreacted monomers will be liberated from the inclusion matrix. The liberated free monomers can be detected by broad line NMR through free rotation of liberated monomers [11, 12]. In spite of the fact that these monomers (VC/AN) are typical electronegative polar monomers equally, quite different configuration is attained (*syndiotactic/VC* and *isotactic/AN*, respectively). The reason why such different microtacticity is attained has neither been clarified experimentally nor theoretically.

3. Experimental

Canal PVC was a white powder and was totally insoluble to organic solvent such as N,N-dimethylformamide (DMF) and dimethyl sulfoxide (DMSO). *Characterization* is given in **Table 1**. It is important to notice that *sample A* means urea canal PVC prepared as described here while *sample B* does free radical one (commercial PVC: Zeon 103, EP-8, straight type/homopolymer). *Sample C* was a standard polymer taken from 100 kinds of standard polymers (Scientific Polymer Products, Inc., USA).

Type	Symbol	Code	Form	Color	Solubility[4]
Canal[1]	*A*	γ-Ray	Powder	White	×
Radical[2]	*B*	Zeon 103	Particle	White	○
Radical[3]	*C*	#038	Powder	White	○

[1]*Prepared in JAERI, Takasaki by urea canal polymerization as described in the text.*
[2]*Supplied from production company, Japan Zeon Co. Ltd. This polymer was a homopolymer.*
[3]*Supplied from the Scientific Polymer Products (USA). This was a high molecular weight homopolymer.*
[4]*Solubility into DMF at room temperature. (0), soluble, (x) insoluble.*

Table 1.
Characterization of PVC samples.

Various measurements were carried out by using instrumental analyses such as solid-state NMR, FT-IR (KBr method), WAXD, TG, solid-state ESR, etc. under carefully controlled conditions. For example, typical NMR conditions in **Figure 7** were as follows: spectrometer, JEOL JNM GX-270; nucleus, ^{13}C; PW2 contact time, 8 ms; PW1 pulse width (90° pulse), 4.5 μs; PD repetition time, 5000 s; number of scans, 150,000; probe, Doty's ceramic probe (7 mm$^{\Phi}$); and external reference, secondary peak of adamantane.

4. Results and discussion

4.1 Characterization of PVC powder

Figure 4 shows IR spectra [12]. There was a distinct difference in the band below 700 cm^{-1} region. Only two peeks are clearly observed in sample A. Lack of the left-hand side band (690 cm^{-1}: assigned to be isotactic in amorphous region) suggests that the urea canal PVC has syndiotactic configuration. It is apparent that Krimm's novel finding was confirmed in this way.

A small difference was also observed in 2900 cm^{-1} region. A new band (2933 cm^{-1}) appeared especially in the urea canal PVC (**Figure 5**). This band may

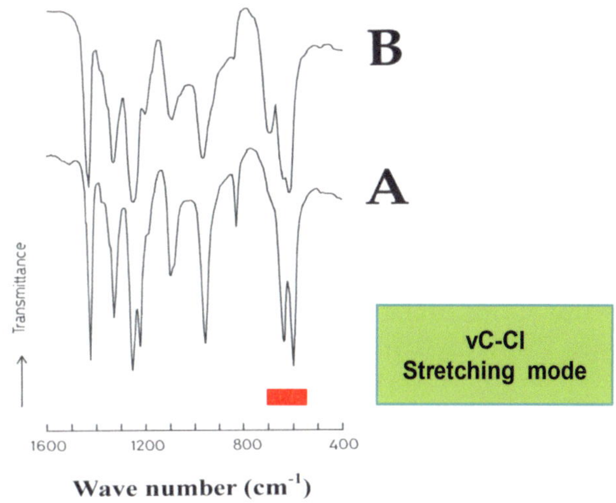

Figure 4.
Comparison of IR spectra of two kinds of PVC.

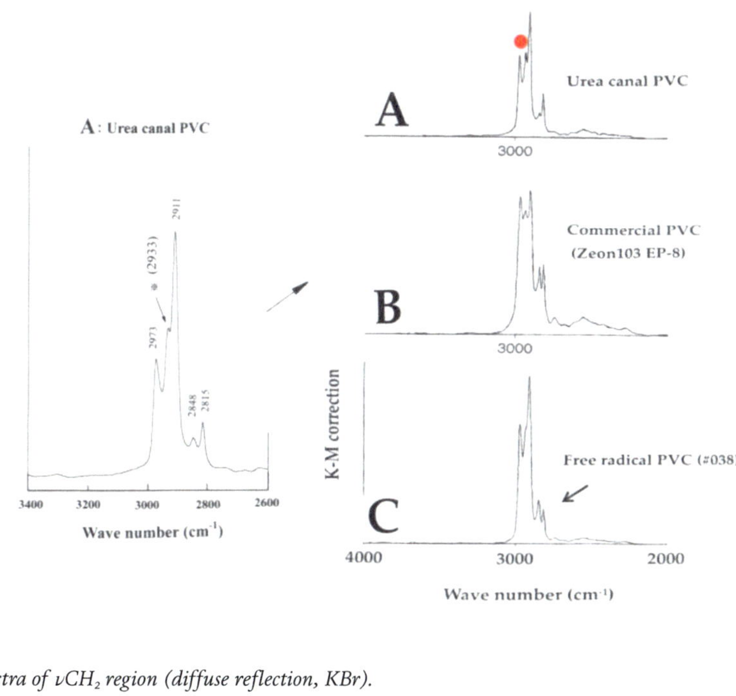

Figure 5.
FT-IR spectra of νCH_2 region (diffuse reflection, KBr).

be correlated with some stereochemical structural factors including *Fermi resonance* and the like, which appears in highly symmetric molecules with very strong intermolecular interaction (enhanced intermolecular hydrogen bonding). The existence of such IR characteristic band has already been pointed out by Tasumi [13]. In fact, this band is directly related to the structurally well-ordered region in PVC sample, since in WAXD results, many Debye-Scherrer rings are clearly observed (see next section).

Figure 6 shows the WAXD results. The existence of many coaxial Debye-Scherrer rings in the sample A is apparent. Careful observation revealed sixth diffraction rings can be counted. With regard to WAXD measurements,

Sakurada et al. already observed sixth diffraction rings in γ-ray PVC (consists of three components indicating partial dissolution into DMF solvent) [14], but they didn't show a WAXD photograph. Even in our experiments, it was very difficult to show rings by ordinary WAXD photograph technically; therefore, we preferred to the direct observation of negative film. By this method, the number of higher order of reflected rings can be counted. Further, very thin but distinct coaxial rings in sample B (commercial one) are obvious.

Figure 7 shows the solid-state NMR spectra. (Since this polymer didn't dissolve in any organic solvent, high-resolution NMR spectra couldn't be obtained.) In comparison with that of free radical PVC (sample B), there was a distinct difference in the NMR spectra. PVC shows two ^{13}C NMR peaks deriving from CH and CH_2 groups on a polymer backbone. Their relative height was almost equal in *sample B*, but it was quite different in *sample A*. The peak area of CH and CH_2 signals in each sample was almost equal, however. From the broadness of NMR spectra, it is possible to consider that three peaks (mm, mr, rr) are separated to some extent toward

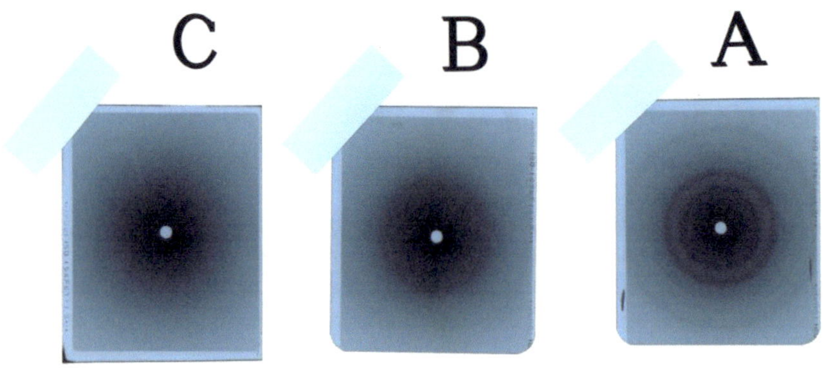

Figure 6.
Comparison of WAXD of three kinds of PVC (negative film).

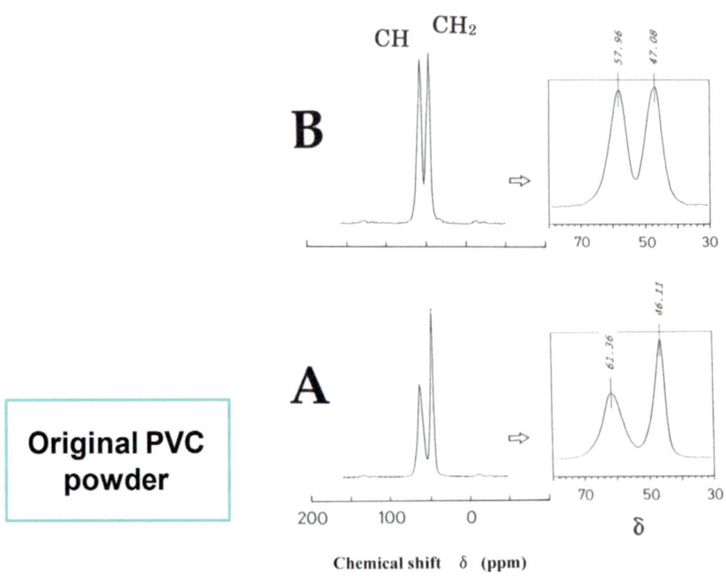

Figure 7.
Comparison of solid-state ^{13}C NMR spectra of two kinds of PVC.

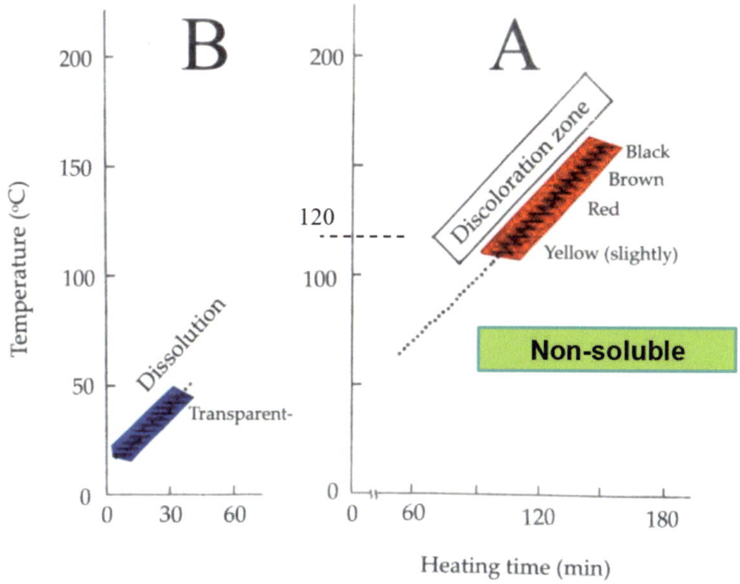

Figure 8.
Solubility of PVC in DMF solvent.

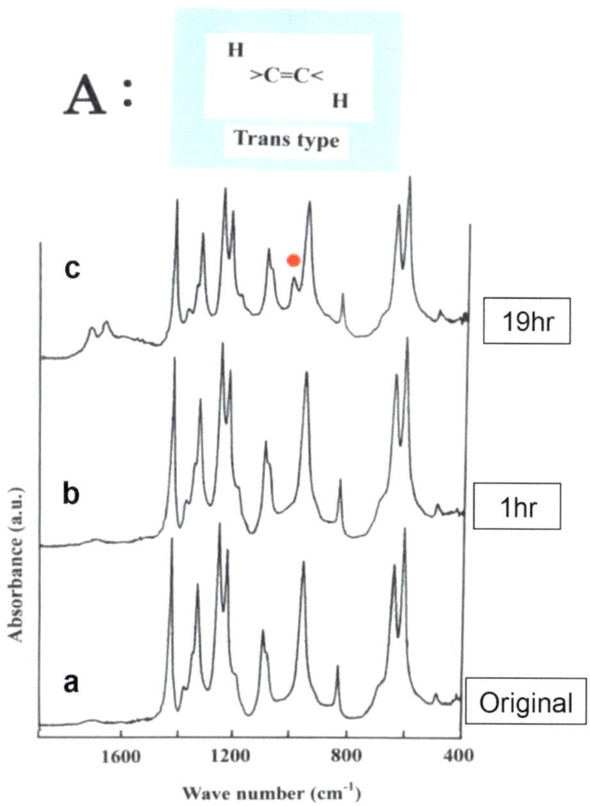

Figure 9.
FT-IR of PVC suspension heated in DMF 1080 cm^{-1} at 120 °C: a: original, b: 1 h, and c: 19 h.

the outer direction and CH peak appeared as a broad single peak with the whole envelope of these peaks [15].

Microtacticity of γ-ray PVC couldn't be determined by solid-state NMR in this way, although quite different NMR peak shape was obtained.

4.2 Solubility in DMF solvent

Figure 8 shows the solubility results of PVC. Free radical PVC (*sample B*) dissolved in DMF completely at room temperature and provided apparently transparent solution. Urea canal PVC (*sample A*), however, didn't. It did a suspension; PVC powder was dispersed in the solvent. Continuous heating up to a high temperature caused a slightly discoloration from white to pale yellow, red, brown, and black. The heated samples were recovered and FT-IR spectra were taken.

Results are given in **Figures 9** and **10**. In *sample A*, change in IR spectra is very little. Polymer backbone is retained even after heating of 19 hours. This is related to the fact that the sample is insoluble and is heated in a suspension state.

A small band appeared at 1080 cm^{-1}, which can be assigned to be trans-type of double bond [16]. In *sample B*, spectral change was large. New band appeared at 1080 cm^{-1} and became progressively large. The decrease of C-Cl band (700–600 cm^{-1}) is obvious. It is known that elimination reaction (HCl) accompanies the formation of trans-type of double bond. This mechanism was common in both samples. Results are summarized in **Figure 11**.

Figure 12 shows the solid-state NMR spectra [17]. In sample A, spectral change was small. In contrast, in sample B, it was large. The appearance of broad peak at about 125–130 ppm is probably due to the formation of double bond, $-(CH{=}CH)_n^-$. In fact, various rubberlike materials having double bond in its molecular structure

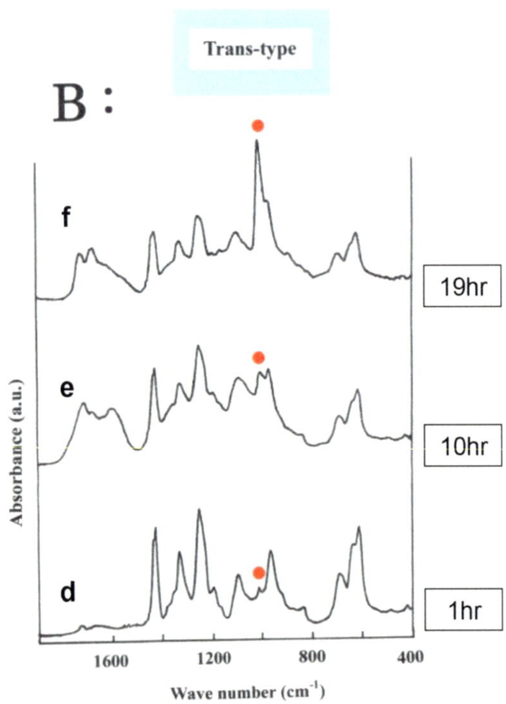

Figure 10.
FT-IR of PVC solution heated in DMF 1080 cm^{-1} at 120°C: d, 1 hour; e, 10 hours; and f, 19 hours.

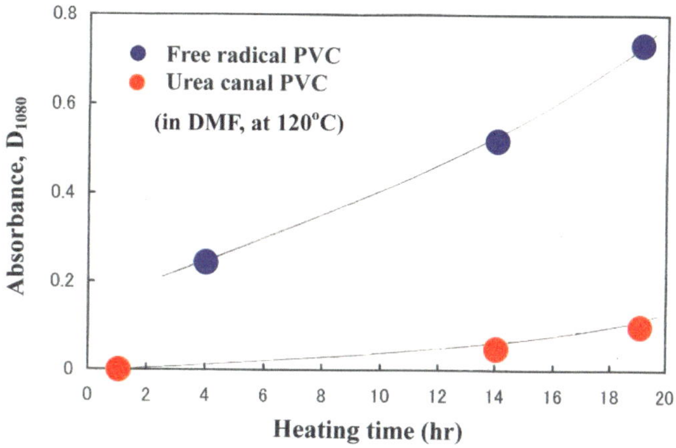

Figure 11.
Variation of IR intensity of trans-type double bond.

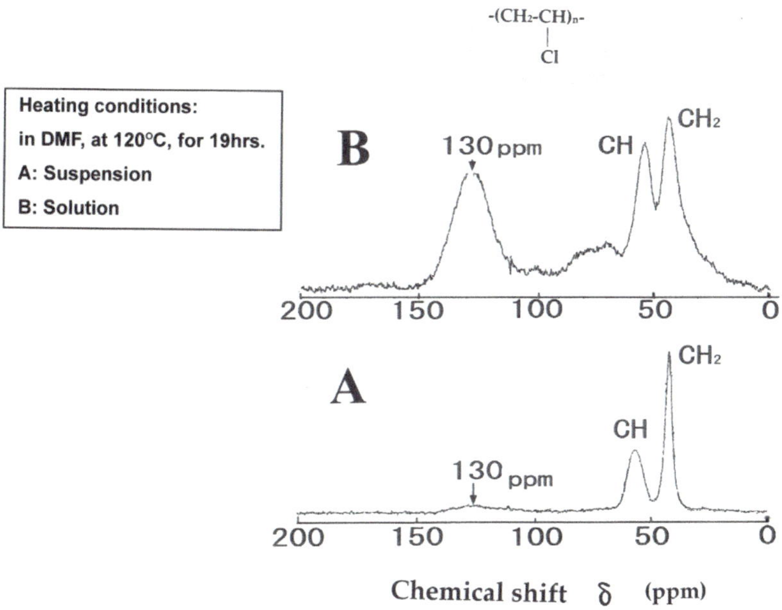

Figure 12.
Comparison of solid-state ^{13}C NMR spectra of heated PVC.

have strong peak in this region [18]. NMR results were as a whole in good agreement with those of FT-IR (**Figures 9** and **10**).

4.3 Discoloration by heat treatment

Figure 13 shows a photograph of heated PVC samples [12]. Canal PVC turns in its color from white (RT) to pale purple (160°C), deep purple (200°C), and then black (~280°C). It is very easy in discoloration. In contrast, free radical one is very hard in its discoloration. One can understand this when compared with the color of two PVC specimens heated up to 200°C (indicated by an arrow). Canal PVC is very deep purple (left), whereas free radical one is only slightly pale orange (right side). Easy discoloration of sample A is probably inherent characteristics of urea canal polymer due to the absence of termination reaction. Because of these characteristics, end radicals are generally living, which would act as a trigger of an elimination reaction (HCl) via a well-known β-elimination mechanism at higher temperatures.

Figure 14 shows solid-state ESR results for sample A under the dynamic heating conditions. Pay attention to the central signal indicated by an asterisk (*), which is derived from the PVC power. Outer several peaks are due to the one from MnO^{+2} inserted as an ESR marker. One can notice that peak intensity increased with the elevation of temperature.

As summarized in **Figure 15**, signal intensification is started at about 160°C, which corresponds to the onset of color changing from white to pale purple (see a

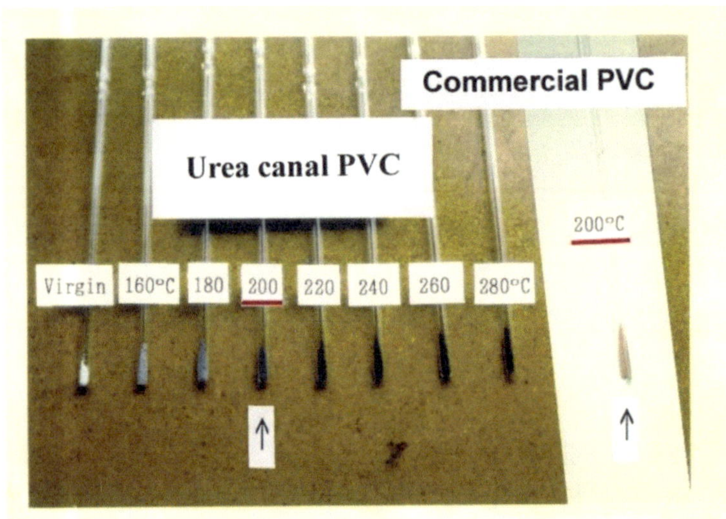

Figure 13.
Comparison of a photograph of heated PVC.

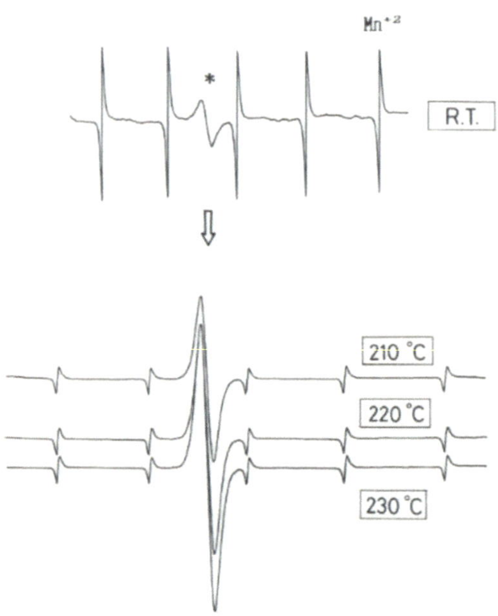

Figure 14.
Solid-state ESR spectra under dynamic heating conditions.

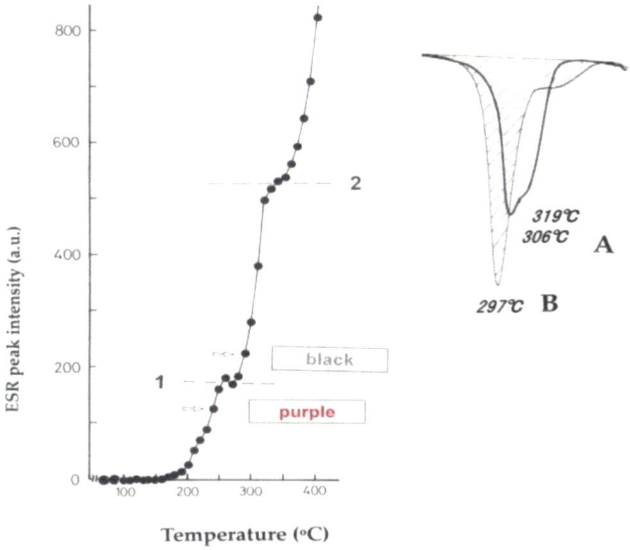

Figure 15.
ESR intensity and TG derivative curve.

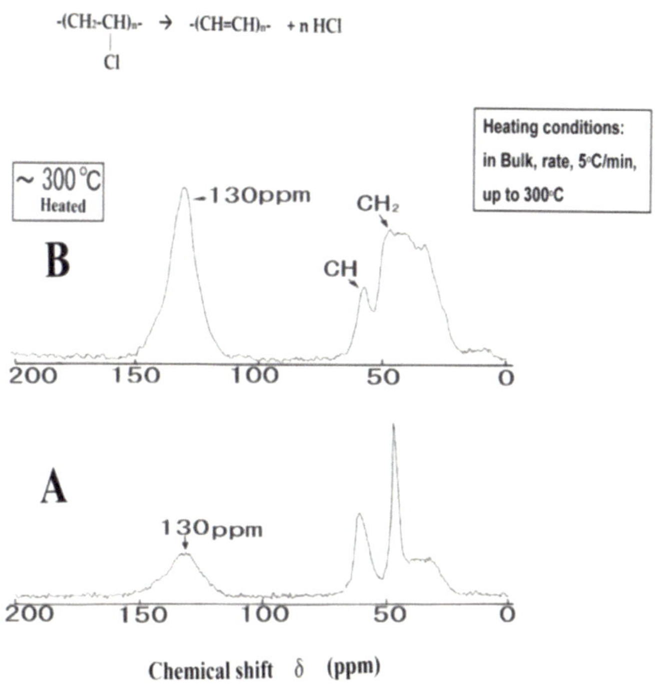

Figure 16.
Comparison of solid-state ^{13}C NMR spectra.

photograph in **Figure 13**). Smooth increase of signal intensity means that elimination reaction proceeds zipper-like (autocatalytically), but its intensity stopped apparently in two regions (*symbols 1 and 2*). An ideal zipper-like reaction started from chain ends in amorphous region was stopped due to *resonance stabilization* and the like, and subsequent reaction may proceed in an irregular form such as interchain elimination or at random fashion type of elimination in unreacted rigid stereoregular sequence.

It is worth while noting that *step 2* in ESR signal agrees with the higher temperature peak of the double peak (306°C–319°C) in TG derivative curve, which were measured in other independent experiments. At any rate, the presence of non-propagating stopping mechanism is obvious. There may be some fine structure and the like in both steps, but these features could not be detected experimentally.

Figure 16 shows solid-state NMR spectra. The original structure has retained considerably in sample A, since CH_2 and CH peaks are clearly present. In contrast, in sample B, the original structural peaks have lost and new broad peak appeared at 130 ppm region. This spectral change in NMR is basically very close to those of FT-IR spectra in **Figures 9** and **10** (heat treatment in the solution or suspension state). In NMR measurements, it should be noted that the peak area is directly proportional to the concentration of functional groups in question; therefore, from both the position (chemical shift) and the peak shape (area), one can understand the mechanism wholly or intuitively.

5. Conclusion

1. The principle of urea clathrate polymerization at low temperatures was described from a purely experimental point of view. The basis lies on the canal complex formation of urea at low temperatures. Geometrical shape and size of the canal complex were described.

2. Characterization of bulk PVC sample was carried out by using FT-IR, WAXD, and solid-state NMR spectra. Lack of isotactic sequence was confirmed by IR spectra of 700–600 cm^{-1} region. Presence of well-ordered region was suggested by FT-IR characteristic band (2933 cm^{-1}), WAXD rings, and solid-state NMR.

3. Solubility into DMF was studied. Free radical sample showed perfect solubility in DMF, while canal sample didn't. Suspension was obtained. Heat treatment caused an elimination reaction followed by formation of trans-type of double bond. The reaction rate of sample A was small.

4. Discoloration by heat treatment was described. Canal PVC showed easy discoloration but rather delay in the TG degradation. The lack of termination reaction is related to the easy discoloration (β-elimination). Slow degradation will be attributed to the stiffness of molecular chain.

5. Solid-state ESR measurements were made. Signal intensity increased exponentially with temperature, but two abrupt stopping regions appeared. The existence of some modification of the mechanism (zipper-like autocatalytic elimination) under uniform heating conditions became apparent.

6. Solid-state NMR was used in various steps such as the characterization of original PVC and thermally degraded samples. Structural change can be visually understood by the appearance of a new peak. Since the peak area is directly proportional to the amount of functional group, a whole understanding of the extent of the progress of degradation is greatly of help.

Acknowledgements

We wish to express our hearty thanks to Mr. H. Sugisawa (JEOL), Prof. Y. Nozawa (Tohoku University), and Mr. T. Katoh (Yamagata University) for NMR measurements. Thanks are also expressed to Prof. M. Matsuda, Prof. T. Miyashita, and Dr. F. Bae (Tohoku University) for many conveniences in solid-state ESR measurements. The valuable discussion on IR spectroscopic data with Prof. M. Tasumi (University of Tokyo) is also greatly acknowledged.

Author details

Masatomo Minagawa[1*], Jun Yatabe[2], Fumio Yoshii[3], Shin Hasegawa[3], Nobuhiro Sato[4] and Tomochika Matsuyama[4]

1 NPO: Dream-Create-Laboratories, Yonezawa, Japan

2 Teikyo University of Science, Uenohara, Japan

3 Takasaki-Establishment, Japan Atomic Energy Research Institute, Takasaki, Japan

4 Research Reactor Institute Kyoto University, Kumatori, Japan

*Address all correspondence to: gakusai-minagawa@memoad.jp

References

[1] Brown JF Jr, White DM. Stereospecific polymerization in thiourea canal complexes. Journal of the American Chemical Society. 1960;**82**:5671-5678

[2] White DM. Stereospecific polymerization in urea canal complexes. Journal of the American Chemical Society. 1960;**82**:5678-5683

[3] Krimm S, Berens AR, Folt VL, Shipman JJ. Assignment of C-Cl stretching modes of polyvinyl chloride. Indian Journal of Chemistry. 1958;**1512**:1513

[4] Krimm S, Berens AR, Folt VL, Shipman JJ. Carbon-chlorine stretching modes of polyvinyl chloride. Chemistry & Industry. 1959;**433**

[5] Shimanouchi T, Tasumu M. Normal coordinate treatment and assignment of IR absorption bands of polyvinyl chloride. Bulletin of the Chemical Society of Japan. 1961;**34**:359-365

[6] Tasumi M, Shimanouchi T. Polarized IR studies on polyvinyl chloride film-Effect of drawing. Spectrochimica Acta. 1961;**17**:731-754

[7] Shimanouchi T. "Vibrational Spectroscopy–Historical Basis and its Chemical Applications", by Memorial Event Group for Celebration of 60 Years Old of Prof. Shimanouchi, T. Tokyo: University of Tokyo; 1977

[8] Tsuruta T. Polymer Synthetic Reactions. Rev. ed. Tokyo: Nikkan Kogyo Shinbun; 1959

[9] Chatani Y. In: Ohtsu T, Takayanagi M, editors. In Progress in Polymer Science, Japan. Tokyo: Kodansha; 1977

[10] Minagawa M, Yamada H, Yamaguchi K, Yoshii F. γ-Ray irradiation canal polymerization conditions ensuring highly stereoregular (>80%) poly(acrylonitrile). Macromolecules. 1980;**25**:503-510

[11] Yoshii F, Abe T, Hayakawa N. Radiation-induced polymerization of vinyl chloride in urea canal complex as studied by broad line NMR. Kobunshi Ronbunshu. 1975;**32**:429-432

[12] Minagawa M, Narisawa I, Sugisawa H, Hasegawa S, Yoshii F. Solid state characterization and the thermal properties of stereoregular poly(vinyl chloride) prepared by urea clathrate polymerization. Journal of Applied Polymer Science. 1999;**74**:2820-2825

[13] Tasumi M. Private Communication

[14] Sakurada I, Nanbu K. On the polymerization by γ–irradiation on VC/urea inclusion compound. Kogyo Kagaku Zasshi. 1960;**81**:1011-1012

[15] Bovey F. Chain Structure and Conformations of Macromolecules. NY: Academic Press; 1982. p. 84

[16] Bellamy LJ. Infrared Spectra of Complex Molecules. 3rd ed. Vol. 1. London: Chapman and Hall; 1975. pp. 38-50

[17] Minagawa M, Katoh T, Yatabe J, Sato N, Matsuyama T. Solubility, discoloration, and ^{13}C NMR spectra of stereoregular poly(vinyl chloride) prepared by urea clathrate polymerization at low temperatures. Preprints of Annual Meeting of the Fiber Society, Shanghai, May 25-27, 2015

[18] Bovey F. Chain Structure and Conformations of Macromolecules. NY: Academic Press; 1982. pp. 115-118

Chapter 7

Aliasing Compromises Staggered-Rotamer Analysis of Polypeptide Sidechain Torsions

Jürgen M. Schmidt

Abstract

Circular undersampling and the ensuing aliasing effect are demonstrated to compromise nuclear magnetic resonance (NMR)-based molecular torsion-angle analysis referring to experimental ^3J-coupling constants when employing the staggered-rotamer model, also known as Pachler model. This popular model is flawed insofar as it systematically produces counterintuitive probabilities for the two minor constituents out of the total three rotamers, to the effect that the apparent circular mean direction of the molecular bond conformation is inflected about its main rotamer angle, a situation that apparently went unnoticed for more than 50 years. The principal reason for systematic errors lay in the model's ill-conceived attempt to resolve the bimodal ^3J-coupling-angle dependency by a mere three discrete points on the circle, thereby conflicting with the Nyquist-Shannon sampling theorem. An anti-aliasing approach is being offered that helps improve the results.

Keywords: circular distribution, directional data, probability density, torsion angle conformation, staggered-rotamer equilibria, discrete Fourier transform, sampling theorem, undersampling, anti-aliasing, vicinal coupling constants, ^3J, aminoacid sidechain, protein structure

1. Introduction

Atom-atom bonds in a molecule often give rise to rotational degrees of freedom, also known as torsion angles. A torsion angle, also known as dihedral angle, is formed by three consecutive bonds in a molecule and defined by the angle between the two outer bonds projected onto a plane perpendicular to the central bond (**Figure 1**).

Finding out how two parts on either side of a rotatable bond relate to each other, that is, assigning a value to the torsion angle, presents one of the challenges in molecular structure determination. A molecule adopting different geometric arrangements—without breaking or making bonds—is said to exhibit distinguishable conformers. Nuclear magnetic resonance (NMR) spectroscopy [2] is uniquely positioned to help characterize not only static molecular structure, but also dynamical processes that involve interconversion between conformers on a short, typically nanosecond timescale. Studying torsion-angle geometry and dynamics by NMR benefits from the measurement of ^3J coupling constants [3]. Typically on the order of a few Hertz, these are magnetic interaction parameters between atoms X and Y in

IntechOpen

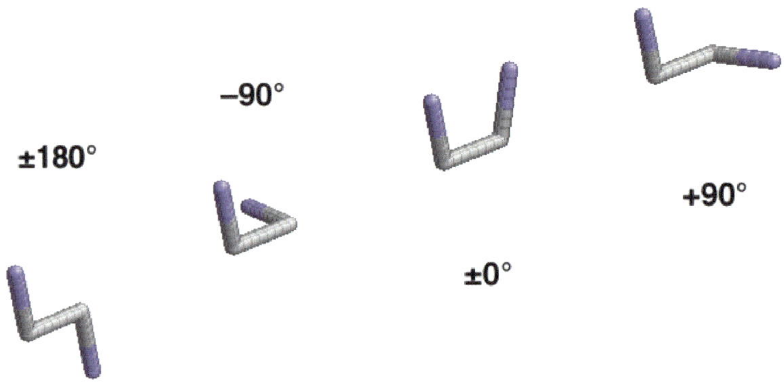

Figure 1.
Basic torsion-angle fragment depicted for various angles of rotation. IUPAC-IUB recommendations [1] assign positive *angles to* clockwise *twists about the central bond of a torsion. Identical torsion-angle sign and value result regardless of viewing down the central bond 'front-to-rear' or 'rear-to-front'.*

the three-bond, four-atom fragment X-A-B-Y constituting dihedral angle θ_{XY} about bond AB. In essence, $^{3}J_{XY}$ coupling interaction is strongest when bonds XA and BY are orientated parallel ($\theta_{XY} = \pm 180°$) and weakest if perpendicular ($\theta_{XY} \sim \pm 90°$). A secondary smaller maximum exists for $\theta_{XY} = \pm 0°$.

1.1 The Karplus curve

Karplus [4] formulated a universal empirical relation as to how a ^{3}J coupling constant depends on the intervening torsion angle:

$$^{3}J_{XY}(\theta) = C_0 + C_1 \cos(\theta + \Delta\theta) + C_2 \cos(2\theta + 2\Delta\theta) \tag{1}$$

Karplus coefficient C_0 signifies the *average* coupling strength obtained for a complete torsion-angle revolution. The C_2 term lends the Karplus curve its *bimodal* shape and determines the undulation depth. The C_1 term adds a *unimodal* component that affects the difference between primary and secondary maxima. Typically, C_1 is negative in order for the primary coupling-constant maximum to appear in the so-called *trans* conformation, when the X-A-B-Y angle is at $\pm 180°$ and the atom-atom interaction is strongest.

We here focus on biomolecular structure analysis of the so-called χ_1 torsion in aminoacid residues, the constituent units of polypeptide and protein chains [1], where the value of χ_1 refers to the dihedral angle $\theta_{N'C\gamma}$ about the bond between carbons C_α and C_β (**Figure 2**). It is imperative to understand, that for each of the up to nine possible atom-pair combinations formed from X = N′, C′, or H^α ("front") and Y = C^γ, $H^{\beta 2}$, or $H^{\beta 3}$ ("rear"), the phase parameter $\Delta\theta$ in the Karplus curve must

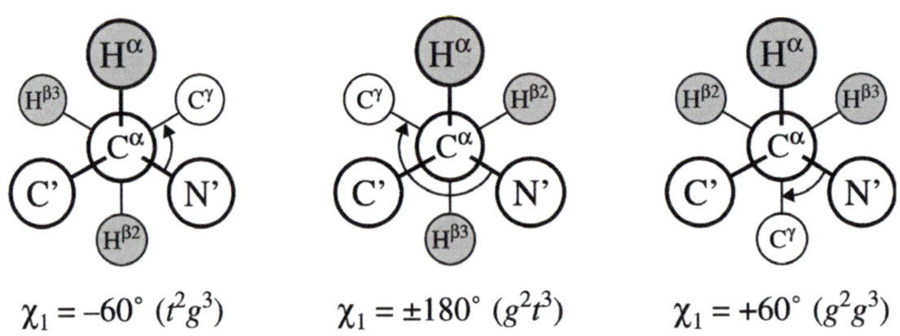

Figure 2.
Staggered-rotamer states for aminoacid sidechain torsion χ_1 viewed down its central C^α—C^β bond. Most aminoacids exhibit two H^β protons oriented trans *or* gauche *with respect to H^α as indicated by a shorthand [5].*

be either 0° or +120° or −120°. This crucially sets up and determines the invariable three-point sampling of the Karplus curve in the situation of χ_1.

1.2 The staggered-rotamer model

Pachler [6] employed $^3J_{H\alpha H\beta 2}/^3J_{H\alpha H\beta 3}$ coupling data pairs in order to quantify populations of conformational states in ethane-like compounds, such as those encountered in aminoacid sidechains. According to Pachler's model, the torsion χ_1 in the aminoacid sidechain is considered to adopt one of three, or to hop between two or between all three so-called *staggered rotamers*, characterized by nominal χ_1 angle values of −60°, ±180°, or +60° (**Figure 2**). An alternative set of angles comprising −120°, ±0°, and +120° would give rise to *eclipsed* rotamers, which, however, are unfavorable in view of molecular energy and, thus, are disregarded for reasons of limited or vanishing lifetime. Both *Pachler model* and *staggered-rotamer model* are synonymous.

The Pachler model allows one to deduce probabilities for each of the three staggered states once a pair of experimental ^3J coupling constants is available. Historically, hydrogen-hydrogen coupling constants $J_{H\alpha H\beta 2}$ and $J_{H\alpha H\beta 3}$ were the first ones accessible to measurement [7], heteronuclear coupling constants were alternatively used later [8]. Accordingly:

$$\begin{aligned} P_{-60} &= \left(J_{H\alpha H\beta 2} - J_{\pm 60^\circ}\right)/\left(J_{180} - J_{\pm 60^\circ}\right) \\ P_{180} &= \left(J_{H\alpha H\beta 3} - J_{\pm 60^\circ}\right)/\left(J_{180} - J_{\pm 60^\circ}\right) \\ P_{+60} &= 1 - P_{-60} - P_{180} \end{aligned} \tag{2}$$

Eq. (2) derives staggered-rotamer probabilities from the measured J_{XY} coupling constants by interpolating between specific coupling values J_{180° and $J_{\pm 60^\circ}$, also designated as J_{trans} and J_{gauche}, respectively, that correspond to the specified fixed geometries for the θ_{XY} angle.

The Pachler model sets up the experimental observable as the corresponding probability-weighted *state-averaged value of J* (indicated by brackets):

$$\langle J \rangle = P_{180} \cdot J_{180} + P_{+60} \cdot J_{+60} + P_{-60} \cdot J_{-60} \tag{3}$$

Both models, the dependency of ^3J couplings on the torsion angle according to Karplus (continuous model) and the dependency of the torsion-angle distribution on relative ^3J coupling-constant pairs according to Pachler (discrete model), form an indispensable basis of many a biomolecular NMR structure investigation.

2. Theory

2.1 Continuous-discrete model interconversion

Sets of probabilities, associated with the three staggered rotamers in the traditional Pachler model of aminoacid sidechain torsion-angle variability, transform into equivalent continuous angle distributions and, vice versa, discrete staggered-rotamer probabilities can be computed from any given angular direction in continuous circular space.

Mean direction θ and concentration R of a continuous angle distribution convert into corresponding normalized probabilities for three discrete-state samples evenly distributed on the circle, according to [9],

$$P_{-60} = (1/3)\left\{1 + R\cos\theta - \sqrt{3}R\sin\theta\right\}$$
$$P_{180} = (1/3)\{1 - 2R\cos\theta\} \quad (4)$$
$$P_{+60} = (1/3)\left\{1 + R\cos\theta + \sqrt{3}R\sin\theta\right\}$$

Conversely, discrete staggered-rotamer probabilities aggregate into the *circular mean direction* of a continuous angle distribution,

$$\tan\theta = \sqrt{3}(P_{+60} - P_{-60})/\{3(P_{+60} + P_{-60}) - 2\} \quad (5)$$

and its *circular order parameter* (concentration) runs between 1 (static fixed angle) and 0 (rotational average),

$$R = \left\{1 + 3\left(P_x^2 + P_y^2 - P_x - P_y + P_xP_y\right)\right\}^{1/2} \quad (6)$$

where P_x and P_y signify any pair out of the three probabilities P_{-60}, P_{180}, and P_{+60}.

2.2 Three-point discrete Fourier transforms

Three-point discrete Fourier transforms (DFT) using kernel, $\omega = \exp.(-i2\pi/N)$, such that a phase of $-i/N$ turns equates to 120° phase rotation if $N = 3$, are expressed in matrix form as [10]

$$\begin{bmatrix} M_0 \\ M_1 \\ M_2 \end{bmatrix} = \begin{bmatrix} 1 & 1 & 1 \\ 1 & \omega^1 & \omega^2 \\ 1 & \omega^2 & \omega^4 \end{bmatrix} \times \begin{bmatrix} m_0 \\ m_1 \\ m_2 \end{bmatrix} \quad (7)$$

Using complex operator $i = (-1)^{1/2}$, transforms rest on conjugate operators $\omega^1 = \omega^+ = -(1/2)\ (1 + i\sqrt{3})$ and $\omega^2 = \omega^- = -(1/2)\ (1 - i\sqrt{3})$, where $\omega^+\omega^- = 1$. Critically, $\omega^4 = \omega^1$ also. Factors (1/2) and $(i\sqrt{3}/2)$ identify with $\cos(\pi/3)$ and $i\sin(\pi/3)$, respectively, such that Eq. (7) recasts as

$$M_0 = m_0 + m_1 + m_2$$
$$M_1 = m_0 - (m_1 + m_2)\cos(\pi/3) - i(m_1 - m_2)\sin(\pi/3) \quad (8)$$
$$M_2 = m_0 - (m_1 + m_2)\cos(\pi/3) + i(m_1 - m_2)\sin(\pi/3)$$

While constituent M_0 represents pure real data, M_1 and M_2 form a complex-conjugate pair sharing identical real parts. Sums, $\cos\theta = (m_1 + m_2)$, and differences, $\sin\theta = i(m_1 - m_2)$, respectively, form the real and imaginary components of the *circular direction* in the continuous-discrete model transformation of Eq. (5).

Swapping ω and its conjugate yields the inverse transform, including normalization, as

$$\begin{bmatrix} m_0 \\ m_1 \\ m_2 \end{bmatrix} = \frac{1}{3}\begin{bmatrix} 1 & 1 & 1 \\ 1 & \omega^2 & \omega^1 \\ 1 & \omega^1 & \omega^2 \end{bmatrix} \times \begin{bmatrix} M_0 \\ M_1 \\ M_2 \end{bmatrix} \quad (9)$$

Now consider probabilities at the $N = 3$ equispaced samples on the circle, that is, $P_{180} = M_0/3$, $P_{-60} = M_1/3$, and $P_{+60} = M_2/3$ inserted in Eq. (9), informing the first three *modes* $p_{m\,=\,0,1,2}$ of a circular probability density, generically given by

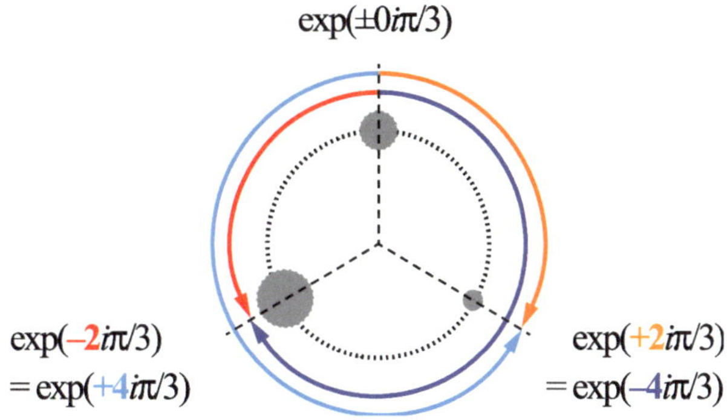

Figure 3.
Directional ambiguity in circular 3-point systems due to degeneracy in circular modes 1 and 2.

$$p_m = N^{-1}\sum_k (\cos m\theta_k - i \sin m\theta_k) \tag{10}$$

Continuous-discrete model interconversion is transparent regarding Fourier transform. Combining circular modes p, the transforms of the P domain, yields the circular mean direction as the *first circular moment*,[1]

$$\tan\theta = (p_1 - p_2)/(p_1 + p_2) \tag{11}$$

and the (squared) circular order parameter, related to auto-correlation [9], as the *second circular moment*,

$$R^2 = p_1 \cdot p_2 \tag{12}$$

The zero-th mode quotes the trivial chance of finding the direction at all,

$$p_0 = 1 = P_{180} + P_{-60} + P_{+60} \tag{13}$$

First and second modes, respectively, then supposedly represent amplitudes of features occurring *once-per-cycle* (m = 1) and *twice-per-cycle* (m = 2),

$$\begin{aligned} p_1 &= P_{180} - (1/2)\left\{P_{-60} + P_{+60} - i\sqrt{3}(P_{-60} - P_{+60})\right\} \\ p_2 &= P_{180} - (1/2)\left\{P_{-60} + P_{+60} + i\sqrt{3}(P_{-60} - P_{+60})\right\} \end{aligned} \tag{14}$$

However, these are conjugates of the *same circular frequency*! Thus, mode p_2 does *not* reflect twice the rate of p_1. This peculiarity is unique to the 3-point DFT and ambiguates the meaning of *single* and *double* phase advances through the circle (**Figure 3**), because if, and only if N = 3, the following identities hold:

$$\exp(\mp i2\pi/3) = \exp(\pm i4\pi/3);\ \cos(\pi/3) = -\cos(2\pi/3);\ \sin(\pi/3) = \sin(2\pi/3) \tag{15}$$

implying $\omega^{+2} = \omega^{-1}$, $\omega^{-2} = \omega^{+1}$.

---

[1] In analogy to linear statistics, circular modes are series of multiple-angle arguments, whereas circular moments are power series in the trigonometric operators. Moments are simple combinations of modes, the inverse does not hold.

Apparently, periodic wrapping equates the double angular speed, ω^2, to the conjugate of the single angular speed, ω^-, thereby folding—under sign inversion—samples M_1 and M_2 onto each other, as visualized in **Figure 4**.

All issues regarding sampling-rate doubling, sign change in the real portion of the complex numbers, and data-pair ambiguity hold for both forward and inverse three-point transforms. Both setups produce identical results, making reconstructing clockwise and counterclockwise arrangements of the circular samples, or P_{+60} and P_{-60}, for that matter, somewhat impossible. Entirely independent of input and output data, the effects arise solely from the transformation operator and are, therefore, *model-inherent*.

Regarding the unique case of three staggered rotamers, let us also consider Karplus coefficients $C_0 = m_0$, $C_1 = m_1$, and $C_2 = m_2$ subjected to three-point DFT according to Eq. (7) and obtain precisely those specific J-coupling values needed to compute the state-averaged J value of Eq. (3):

$$\begin{aligned} J_{trans} &= J_{180} = C_0 - C_1 + C_2 \\ J_{gauche} &= J_{\pm 60} = C_0 + C_1/2 - C_2/2 \end{aligned} \tag{16}$$

Inverse-transforming the values of J_{trans} and J_{gauche}, inserted as $J_{180} = M_0/3$, $J_{+60} = M_1/3$, and $J_{-60} = M_2/3$ into Eq. (9), recovers the three coefficients (i.e., modes) of the Karplus Eq. (1).

Mathematically, this looks clean and entirely reversible. In analogy to the modes of the circular probability density in Eq. (14), the DFT obtained for both J_{+60} and J_{-60} inserted in Eq. (3) would necessarily result in complex-conjugate numbers. And yet, deliberately ignored was the fact that theoretical as well as experimentally

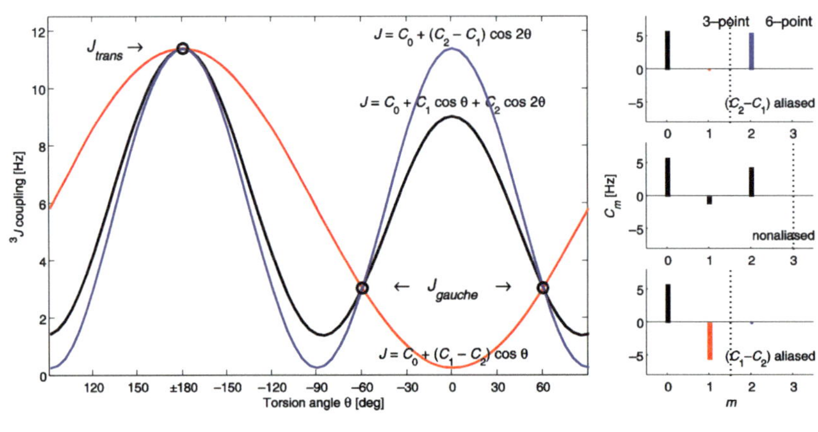

Figure 4.
Main panel: effective Karplus-curve manifold generated by aliasing cosθ and cos 2θ modes. Typical coefficients (C_0, C_1, C_2) and a phase offset of $\Delta\theta = 0°$ apply to $^3J_{H\alpha H\beta_3}$ in aminoacid sidechains. The original (non-aliased) Karplus curve is shown in black. Aliased unimodal (red) and bimodal (blue) curves, respectively, result from coefficient C_2 folding onto C_1 and, vice versa, C_1 folding onto C_2. Associated with staggered rotamers are those invariant focal points J_{trans} and J_{gauche} through which all curves pass. Small panels: sampling properties of rotamer models in contexts of 3-point and 6-point discrete Fourier transforms of Karplus curves. Each coefficient C_m as the mth circular mode (bars) connects with an effective circular sampling rate in units of $(2\pi\ \mathrm{rad})^{-1}$, m being the periodicity of that component. Bottom: sampling the bimodal coupling-angle dependency at three 120°-equispaced (e.g., staggered-rotamer) angles corresponds to rate 1.5 π^{-1} rad^{-1} (dotted line), too low to resolve the higher C_2 mode. Consequently, C_2 inverts and folds into the lower frequency band and stacks onto or aliases the negative C_1 mode (red bar). The original C_2 amplitude appears blanked. Reconstructed from such distorted coefficients, the mis-sampled Karplus curve appears purely unimodal. Top: alternatively, C_1 folds under sign inversion onto the C_2 mode (blue bar), generating a purely bimodal curve. Middle: aliasing of the high mode is avoided with a 6-state staggered-eclipsed 60°-equispaced rotamer model that samples the circle at twice the rate of 3 π^{-1} rad^{-1} (dash-dotted line).

observed J-coupling constants are generally considered real data and, therefore, J values for *gauche*$^+$ and *gauche*$^-$ states are ordinarily identical. Clockwise and counterclockwise orientations become indistinguishable once again. At any rate, disregarding imaginary components amounts to loss of information, introducing a subtle irreversibility into the staggered-rotamer analysis that cannot be corrected.

3. Results

3.1 Worked example: selected sidechain rotamers in a protein

In the course of extensive protein-structure studies by NMR, ^3J coupling constants related to aminoacid sidechain torsions were measured in a variety of proteins. Ongoing investigations target sidechain χ_1 torsion structure in ribonuclease T_1 (RNase T_1), an enzyme of 104 aminoacids size, experimental ^3J coupling data of which are deposited with the BMRB [11]. Applying up to nine available coupling constants as experimental constraints, two models, a continuous single-state torsion angle and the discrete staggered-rotamer populations, were least-squares fitted for each aminoacid in the enzyme. Exemplifying the present issues, RNase T_1 contains three histidines, each of which adopts a different predominant staggered state with only little dispersion about their respective circular mean direction. Accordingly, the torsions lock χ_1 conformations near $\pm 180°$ in His27 (as evident from a large *trans*-$^3J_{\mathrm{N'C\gamma}}$ coupling constant, **Figure 2**), near +60° in His40° (large *trans*-$^3J_{\mathrm{H\alpha C\gamma}}$), and near $-60°$ in His92 (large *trans*-$^3J_{\mathrm{C'C\gamma}}$).

The predominant χ_1 rotamer in His92 is populated at P_{-60} = 92%, with only negligible or minor contributions from P_{180} = 0% and P_{+60} = 8%, which together make for a circular mean direction of $-56°$ (**Figure 5**). The single-state model converges at $-72°$, somewhat "mirrored" or inflected about the ideal $-60°$ staggered state.

The His27 torsion noticeably tilts away from the ideal $\pm 180°$ staggered angle, circular mean direction from the discrete model being $-160°$. However, a χ_1 value of +158° results from fitting a continuous single-state angle and would infer a primary $\pm 180°$ and a secondary +60° rotamer, broadly at 3:1 proportion [9]. Yet, populations emerging from staggered-rotamer analysis (P_{-60} = 23%, P_{180} = 89%, P_{+60} = -12%) suggest $-60°$ as the more significant secondary constituent. This, together with a decidedly negative probability for the +60° rotamer, results in an "opposite" circular mean direction, again inflected about the constituent $\pm 180°$ state.

Similarly, discrete staggered-state populations in His40 emerge as P_{-60} = 0%, P_{180} = 22%, and P_{+60} = 78%, making for a circular mean direction of approximately +76°, while χ_1 converges at +36° in the continuous model. In the staggered-rotamer model, a mean direction deviating from +60° toward smaller angle values would command a significant contribution from the $-60°$ state to the overall average. Once again, the discrete model suggests the less plausible $\pm 180°$ conformer.

Even though both fitted models agree as regards the main conformer in each case, state populations of the minor conformers in the staggered-rotamer model seem always somewhat counterintuitive. Deviating from a single ideal staggered state, the inflection phenomenon in the discrete model consistently manifests as a false apparent mean direction in the rotationally opposite sense, off in terms of degrees by—approximately—as much as the *correct* mean direction derived from the continuous model deviates from that nearest ideal staggered state. How do we know it is the staggered-rotamer model delivering wrong results and not the continuous model? Because were the angle tilted to the "opposite/other side," the

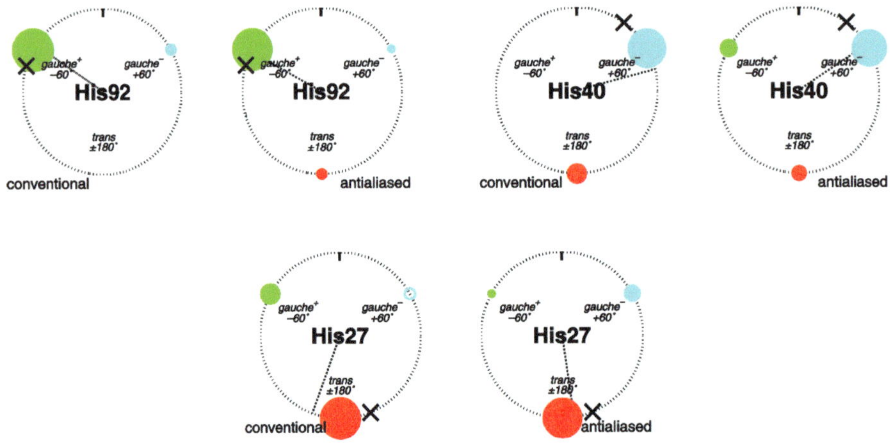

Figure 5.
Conventionally optimized discrete staggered-rotamer probabilities for histidines in RNase T_1, and the effect of anti-aliasing applied to these. Probabilities $P_{\pm180}$ (red), P_{-60} (green), and P_{+60} (blue) shown as area-proportional circles, and dials indicate the effective circular mean direction obtained from these. Crosses mark torsion angles that best fit the J-coupling data using a continuous variable-angle single-state model. Anti-aliasing improves the probabilities insofar as apparent mean directions for the modified sets are closer to a staggered state than for the original and also agree better with torsion angles from the single-state fit result. Anti-aliasing helps inflect the apparent mean direction about the nearest staggered angle toward the opposite side. Most noticeably, His27 is an example of a predominant trans rotamer with clockwise deflection of its apparent circular mean direction, chiefly due to the conventionally fitted negative probability of the +60° state (open circle). Anti-aliasing reverts the apparent mean into a counterclockwise direction and also inverts the implausible value of P_{+60}.

corresponding set of probabilities (top panel of **Figure 6**) would generate a value of *J* typically in disagreement with the experimental value, for example, for $^3J_{H\alpha H\beta 3}$, as read off the Karplus curve for the given angle (bottom panel of **Figure 6**).

Such discrepancy is not limited to specific aminoacid residue types; rather, it is seen—without exception—in all rotamer studies. Revisiting ^3J data collected for a different protein, *Desulfovibrio vulgaris* flavodoxin [12], confirmed suspicion that a systematic issue would be at work. Invariably, counterintuitive admixtures of rotamers are calculated for all non-ideally staggered sidechain conformations.

3.2 A compromise fix: anti-aliasing

Anti-aliasing signifies any procedure that attempts to ameliorate adverse effects from aliasing due to coarse sampling. Anti-aliasing applied in, for example, image processing helps smoothen jagged lines and edges that result from digitizing continuum data into discrete samples [13]. Generally, such approaches form weighted averages over a number of adjacent data points and, therefore, are irreversible data manipulation.

Anti-aliasing applied to the discrete rotamer model helps restore some bimodal curve feature in the distorted coupling angle relationship and would also minimize unrealistic negative excursions in the probability parameters. Considering that Eq. (2) introduces into the traditional interpretation of staggered rotamers a strict one-to-one correspondence between each coupling and precisely one, and only one, staggered-state population, this permits the remaining two population parameters to be manipulated at will without interfering with that respective correspondence, offering an opportunity to diminish differential probability and improve the population parameters generally. The approach taken here represents the improved symmetrical, balanced variant of the range-bound probability fitting devised in [9] by manipulating all probability parameters equally and simultaneously:

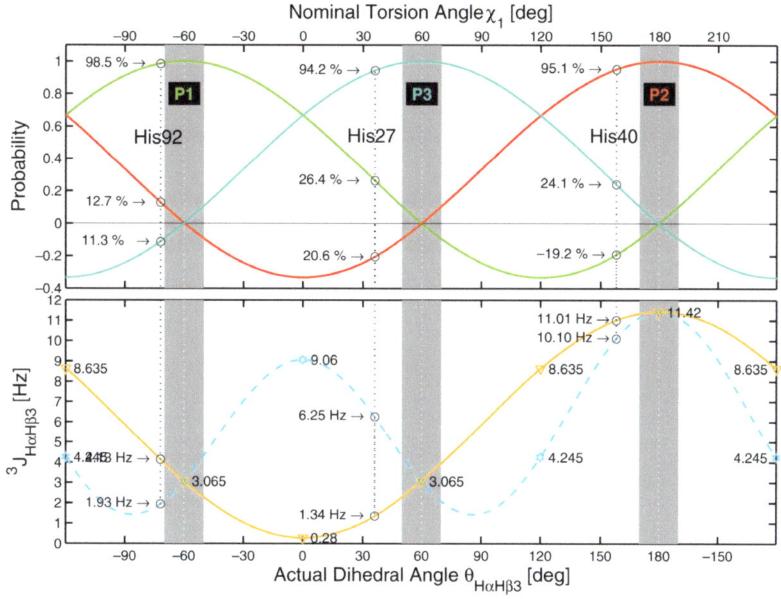

Figure 6.
Top: concerted change in staggered-rotamer probabilities P with varying circular mean direction for torsion χ_1 *according to Eq. (2); bottom: state-averaged aminoacid* $^3J_{H\alpha H\beta 3}$ *coupling derived from these probabilities when aggregated according to Eq. (3). The decidedly unimodal constructed curve (solid line) tracks exclusively the trans rotamer probability* P_2*, thus, failing to reproduce the bimodal angle dependency of J in direct response to an angle sweep according to the original Karplus curve (dashed line) of Eq. (1). Annotated RNase-histidine examples highlight large differential probability between the minor rotamer pairs (top) as well as counterintuitive coupling values (bottom) as seen for all deflections from ideal staggered states (shaded zones).*

$$
\begin{aligned}
{}^{aa}P_{-60} &= P_{-60} + z\left(-P_{-60}P_{180}/2 - P_{-60}P_{+60}/2 + P_{180}P_{+60}\right)\\
{}^{aa}P_{180} &= P_{-60} + z\left(-P_{-60}P_{180}/2 + P_{-60}P_{+60} - P_{180}P_{+60}/2\right)\\
{}^{aa}P_{+60} &= P_{+60} + z\left(+P_{-60}P_{180} - P_{-60}P_{+60}/2 - P_{180}P_{+60}/2\right)
\end{aligned}
\tag{17}
$$

Underlying the anti-aliasing principle in Eq. (17) is the observation that the smallest probability derived from the Pachler model is always too small, if not negative, while that of the main rotamer is always too large. Therefore the small *P* value is being raised (one positive increment) at the expense of both other parameters, from which half the amount each is being taken off (two negative half increments). Applying the correction in turn to each set gives rise to the anti-aliasing matrix in Eq. (17). Normalization is critically ensured as each product term added (or subtracted) is being subtracted (or added) elsewhere.

Values between 0.5 and 1.5 for the anti-aliasing parameter z range form a reasonable compromise between such diverse effects as differential probability, negative probability, and unimodality of the coupling-angle dependence [9]. By choosing $z = 1$, all eclipsed-state populations amount to one-third, making these situations indistinguishable from the complete rotational-average limit. Contrasting the conventional staggered-rotamer model, the anti-aliased model typically inverts and reduces differential probability between the two minor rotamers (**Figure** 7).

Aliasing affects primarily the probabilities, yet, consequently also any back-calculated—as opposed to experimental—coupling values. Only the mean conformation-averaged calculated *J* couplings remain invariant when inserting anti-aliased probabilities ^{aa}P into Eq. (3), equaling those obtained with the unmodified *P* values.

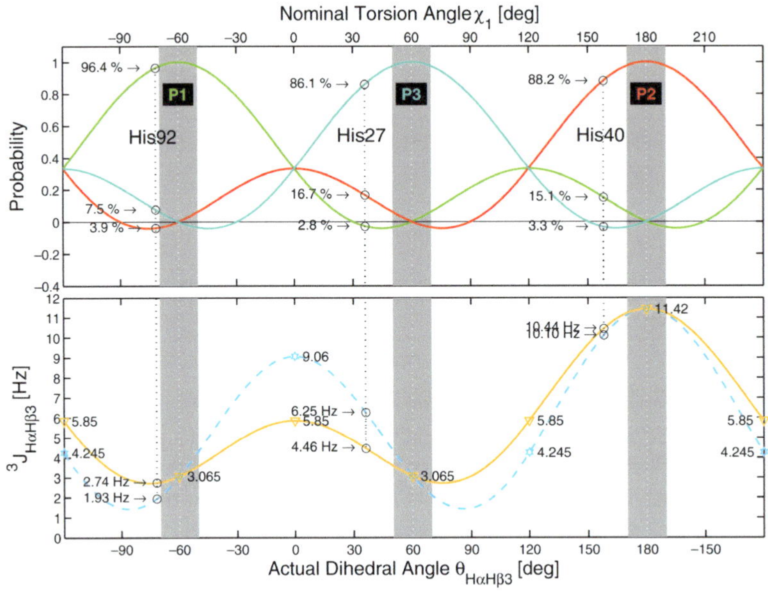

Figure 7.
*Top: staggered-rotamer probabilities for an angle sweep anti-aliased using z = 1 in Eq. (17); bottom: dependency of the state-averaged $^3J_{H\alpha H\beta 3}$ coupling according to Eq. (3) when using anti-aliased staggered-rotamer probabilities, re-introducing into the coupling-angle dependency a certain bimodal feature (solid line), approaching that in the original Karplus curve (dashed line). Other graph features as described in **Figure 6** legend.*

In an attempt to improve the coupling estimates, anti-aliasing can be implemented in two possible ways:

i. Applied as post-processing procedure, the fitted probabilities are being modified (anti-aliased) according to Eq. (17), and coupling estimates in Eq. (3) amended in accordance with the updated values of $P_{x,y,z}$. As least-squares parameter optimization already minimizes the discrepancy between observed and calculated J coupling values, any subsequent parameter modification will invariably increase the overall fit error in J. After all, we are interested less in the fit error than we are in plausible and acceptable molecular descriptions.

ii. Applied during data fitting, anti-aliasing and optimization of the probability parameters will interfere and cancel somewhat. Again, optimization strives to minimize the difference between observed and calculated J values; only this time, the instantaneous fit error feeds back on the continual probability-parameter adjustment, counteracting the anti-aliasing due to the model-inherent divergence between calculated and true coupling-angle dependence.

In practice, parameters optimized following option (ii) converged nearer to those obtained in the original, aliased fit, while more significant corrective effects are seen when adopting option (i).

As circular folding tends to equalize the meaning of both first and second mode coefficients of the Karplus curve, one might equally (or alternatively) employ jointly the C_2 coefficient with an additional folded (mirrored and inverted) C_1 coefficient. This would connect with twice the angular speed, which happens to coincide with the single speed of opposite sign (Eq. (14)). This condition is identical to anti-aliasing using $z = 2$.

Anti-aliasing is irreversible, as the sequence order in two recursions, $z = +1/-1$ vs. $-1/+1$, does make a difference. The latter results in two equal J_{trans} maxima, much like applying $z = +2$ once, and four J_{gauche} values for all other 60° intervals. The probabilities are more peaked at the top and flatter at the bottom. The former sequence produces flat tops in both P and J profiles and peaked bottoms, even though J_{trans} and J_{gauche} are correctly reproduced, and all eclipsed states show J_{mean}.

As multiple couplings are considered in real applications, the final outcome is a balanced compromise between all contributing parts.

4. Discussion and conclusion

In principle, conversion of data into a model encounters one of four possible situations (**Figure 8**), depending on the ratio between the number of independent experimental observations and the number of model parameters to determine. Focusing here on aminoacid geometry:

i. Normally, if many data are available and accurate, the analysis outcome is most likely reliable, as is the case with the polypeptide mainchain ϕ torsion angle, where six J-coupling constants were collected, associated with phases densely spaced at 60° intervals on the circle, allowing to determine the torsion-angle parameter with high accuracy [14].

ii. A number of model variables that exceeds the number of experimental observables renders analysis generally impossible, owing to too few data.

iii. At most, interpretation of insufficient data would produce some artificial result. Indications are that this might be the case with polypeptide mainchain ψ torsion analysis, where a mere three J-coupling constants are accessible for sampling rotamer probabilities associated with three phases at 120° intervals, similar to the present work, yet one more experimental observable to refer to.

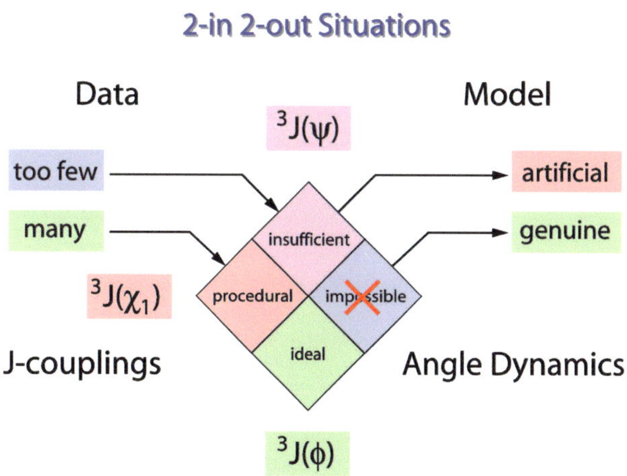

Figure 8.
Possible pathways for conversion of experimental data into abstract or theoretical model parameters.

iv. Finally, the present case of polypeptide sidechain χ_1 torsion analysis would not normally suffer any data shortage in determining a single angle parameter, given a principal theoretical parameter over-determination from up to nine experimental observables. Rather, the analysis procedure itself produces the notoriously distorted artificial outcome. This work demonstrated the mathematical reasoning as to why implausible probabilities are being obtained regardless of an overwhelming data supply.

Attempting to extract meaningful parameters of mean direction and dispersion by digitizing the bimodal *J*-coupling angle dependency through a "bottleneck" of only three staggered-rotamer angles must appear error-prone. In many ways, this seems tantamount to the proverbial "square peg in a round hole." It is hoped that the issues and pitfalls connected with circular three-point sampling and transforms, demonstrated here in a biochemical context, would inspire a fresh look at applications in other disciplines in mathematics, physics, or engineering.

Acknowledgements

Frank Löhr (Goethe-Universität Frankfurt am Main) is thanked for stimulating discussion and his continual interest in *J*-coupling studies.

Author details

Jürgen M. Schmidt
School of Biosciences, University of Kent, Canterbury, United Kingdom

*Address all correspondence to: jmschmidt@btconnect.com

References

[1] IUPAC-IUB commision on biochemical nomenclature. Journal of Molecular Biology. 1970;**52**:1-17

[2] Evans JNS. Biomolecular NMR Spectroscopy. 1st ed. Oxford: Oxford University Press; 1995

[3] Schmidt JM, Löhr F. Refinement of protein tertiary structure by using spin-spin coupling constants from nuclear magnetic resonance measurements. In: Faraggi E, editor. Protein Structure. Rijeka, Croatia: Intech; 2012. pp. 95-120. ISBN: 978-953-51-0555-8. Available from: http://www.intechopen.com/books/protein-structure

[4] Karplus M. Vicinal proton coupling in nuclear magnetic resonance. Journal of the American Chemical Society. 1963; **85**:2870-2871

[5] Wagner G, Braun W, Havel TF, Schaumann T, Go N, Wüthrich K. Protein structures in solution by nuclear magnetic resonance and distance geometry—The polypeptide fold of the BPTI determined using two different algorithms, DISGEO and DISMAN. Journal of Molecular Biology. 1987;**196**: 611-639

[6] Pachler KGR. Nuclear magnetic resonance study of some α-amino acids—II. Rotational isomerism. Spectrochimica Acta. 1964;**20**:581-587

[7] Nagayama K, Wüthrich K. Structural interpretation of vicinal proton-proton coupling constants $^3J_{H\alpha H\beta}$ in the basic pancreatic trypsin inhibitor measured by two-dimensional *J*-resolved NMR spectroscopy. European Journal of Biochemistry. 1981;**115**:653-657

[8] Kessler K, Griesinger C, Wagner K. Peptide conformations 42. Conformation of side chains in peptides using heteronuclear coupling constants obtained by two-dimensional NMR spectroscopy. Journal of the American Chemical Society. 1987;**109**:6927-6933

[9] Schmidt JM. Transforming between discrete and continuous angle distribution models: Application to protein χ_1 torsions. Journal of Biomolecular NMR. 2012;**54**:97-114

[10] Bracewell RN. The Fourier Transform and its Applications. 2nd International ed. New York: McGraw-Hill; 1986

[11] Schmidt JM, Löhr F. ^3J coupling constants related to chi$_1$-torsions in Ribonuclease T$_1$ from Aspergillus oryzae. Biological Magnetic Resonance Data Bank. BMRB Entry 26067 pending release on publication. Available from: http://www.bmrb.wisc.edu/data_library/summary/index.php?bmrbId=26067

[12] Pérez C, Löhr F, Rüterjans H, Schmidt JM. Self-consistent Karplus parametrization of ^3J couplings depending on the polypeptide side-chain torsion χ_1. Journal of the American Chemical Society. 2001;**123**:7081-7093

[13] Foley JD, van Dam A, Feiner SK, Hughes JF, Phillips RL. Introduction to Computer Graphics. 13th Reprint. Boston: Addison-Wesley Publishing Company Inc.; 1997

[14] Schmidt JM, Blümel M, Löhr F, Rüterjans H. Self-consistent ^3J coupling analysis for the joint calibration of Karplus coefficients and ϕ-torsion angles. Journal of Biomolecular NMR. 1999;**14**:1-12